THE WORLD BOOK ENCYCLOPEDIA OF
PEOPLE AND PLACES

1
A-B

a Scott Fetzer company
Chicago
www.worldbookonline.com

For information about other World Book publications,
visit our website at http://www.worldbookonline.com
or call 1-800-WORLDBK (1-800-967-5325).

For information about sales to schools and libraries, call
1-800-975-3250 (United States);
1-800-837-5365 (Canada).

The World Book Encyclopedia of People and Places,
second edition

Library of Congress Cataloging-in-Publication Data

The World Book encyclopedia of people and places.
  v. cm.
  Summary: "A 7-volume illustrated, alphabetically arranged
set that presents profiles of individual nations and other
political/geographical units, including an overview of history,
geography, economy, people, culture, and government of each.
Includes a history of the settlement of each world region
based on archaeological findings; a cumulative index; and Web
resources"--Provided by publisher.
  Includes index.
  ISBN 978-0-7166-3758-5
  1. Encyclopedias and dictionaries. 2. Geography--
Encyclopedias. I. World Book, Inc. Title: Encyclopedia of
people and places.
  AE5.W563 2011
  030--dc22
                                          2010011919
This edition ISBN: 978-0-7166-3760-8

Printed in Hong Kong by Toppan Printing Co. (H.K.) LTD
3rd printing, revised, August 2012

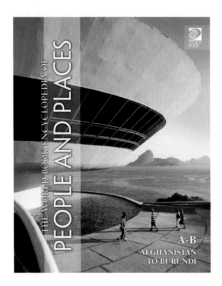

Cover image:
View across Guanabara
Bay to Rio de Janeiro,
Brazil.

© Robert Harding Picture
Library/SuperStock

CONTENTS

HOW TO USE THIS SERIES

There are seven volumes in *The World Book Encyclopedia of People and Places:* Volume 1, A-B; Volume 2, C-F; Volume 3, G-J; Volume 4, K-N; Volume 5, O-S; Volume 6, T-Z; and Volume 7, People on the Move, the cumulative index, and website links for all countries.

Each book presents individual nations and other political or geographic units through articles that are at least two pages long. The countries are arranged in alphabetical order, beginning with Afghanistan in Volume 1 and ending with Zimbabwe in Volume 6. Volume 7 provides a history of the settlement of the world's regions based on archaeological findings.

Articles about a specific place provide an overview of its history, geography, economy, people, culture, and current political situation. You will also find descriptions of the features that make that place unique.

Articles for countries with large populations or great prominence in world affairs generally cover several pages. In such cases, the article is divided into sections covering a particular topic relating to that country, such as History,

Environment, People, and Economy. These sections are always covered within a single two-page spread. An example of an article with several sections is shown below.

A look at the articles

Coverage of each country includes a section that features a physical/political map and fact box. The fact box provides information about government, economy, and people. For example, life expectancy and languages spoken are given, as are birth rate and currency.

Another feature of this series is the timelines. Timelines are found in articles that include a history section. They list significant events in a country's past.

In addition to the physical/political maps, a variety of thematic maps are included in the set. Diagrams and charts provide other information in graphic formats.

A comprehensive index

All the articles are referenced in the cumulative index at the end of Volume 7. The index is an invaluable aid in finding specific information quickly.

Article titles are highlighted in a colored bar.

Photographs in full color complement the text.

Section titles provide quick reference points.

Flags are shown for each country.

Fact boxes contain key information for quick look-ups.

Locator maps show the location of a country on the globe.

Timelines list significant dates in a country's history.

Thematic and **historical maps** provide additional information.

Captions explain the photographs and illustrations and provide additional information.

THE CANADIAN WILDERNESS
CANADIAN CITIES
ECONOMY AND RESOURCES
CONFLICT AND OPPORTUNITY
HISTORY
THE FIRST CANADIANS
ENVIRONMENT
CANADA TODAY

CANADA

23

The article on Canada in Volume 2 contains nine sections.

Illustrations and **diagrams** provide detailed information on how things look or work.

Physical/political maps show country boundaries, names of the surrounding countries, and major features and cities.

Running feet reinforce the name of the article.

POLITICAL WORLD MAP

The world has 196 independent countries and about 50 dependencies. An independent country controls its own affairs. Dependencies are controlled in some way by independent countries. In most cases, an independent country is responsible for the dependency's foreign relations and defense, and some of the dependency's local affairs. However, many dependencies have complete control of their local affairs.

By 2010, the world's population was nearly 7 billion. Almost all of the world's people live in independent countries. Only about 13 million people live in dependencies.

Some regions of the world, including Antarctica and certain desert areas, have no permanent population. The most densely populated regions of the world are in Europe and in southern and eastern Asia. The world's largest country in terms of population is China, which has more than 1.3 billion people. The independent country with the smallest population is Vatican City, with only about 830 people. Vatican City, covering only 1/6 square mile (0.4 square kilometer), is also the smallest in terms of size. The world's largest nation in terms of area is Russia, which covers 6,601,669 square miles (17,098,242 square kilometers).

Every nation depends on other nations in some way. The interdependence of the entire world and its peoples is called *globalism*. Nations trade with one another to earn money and to obtain manufactured goods or the natural resources that they lack. Nations with similar interests and political beliefs may pledge to support one another in case of war. Developed countries provide developing nations with financial aid and technical assistance. Such aid strengthens trade as well as defense ties.

6

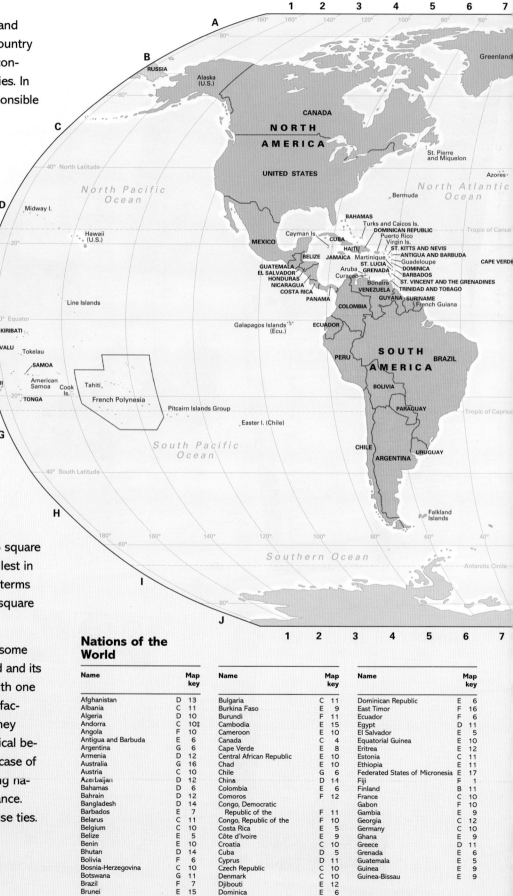

Nations of the World

Name	Map key	Name	Map key	Name	Map key
Afghanistan	D 13	Bulgaria	C 11	Dominican Republic	E 6
Albania	C 11	Burkina Faso	E 9	East Timor	F 16
Algeria	D 10	Burundi	F 11	Ecuador	F 6
Andorra	C 10‡	Cambodia	E 15	Egypt	D 11
Angola	F 10	Cameroon	E 10	El Salvador	E 5
Antigua and Barbuda	E 6	Canada	C 4	Equatorial Guinea	E 10
Argentina	G 6	Cape Verde	E 8	Eritrea	E 12
Armenia	D 12	Central African Republic	E 10	Estonia	C 11
Australia	G 16	Chad	E 10	Ethiopia	E 11
Austria	C 10	Chile	G 6	Federated States of Micronesia	E 17
Azerbaijan	D 12	China	D 14	Fiji	F 1
Bahamas	D 6	Colombia	E 6	Finland	B 11
Bahrain	D 12	Comoros	F 12	France	C 10
Bangladesh	D 14	Congo, Democratic Republic of the	F 11	Gabon	F 10
Barbados	E 7	Congo, Republic of the	F 10	Gambia	E 9
Belarus	C 11	Costa Rica	E 5	Georgia	C 12
Belgium	C 10	Côte d'Ivoire	E 9	Germany	C 10
Belize	E 5	Croatia	C 10	Ghana	E 9
Benin	E 10	Cuba	D 5	Greece	D 11
Bhutan	D 14	Cyprus	D 11	Grenada	E 6
Bolivia	F 6	Czech Republic	C 10	Guatemala	E 5
Bosnia-Herzegovina	C 10	Denmark	C 10	Guinea	E 9
Botswana	G 11	Djibouti	E 12	Guinea-Bissau	E 9
Brazil	F 7	Dominica	E 6		
Brunei	E 15				

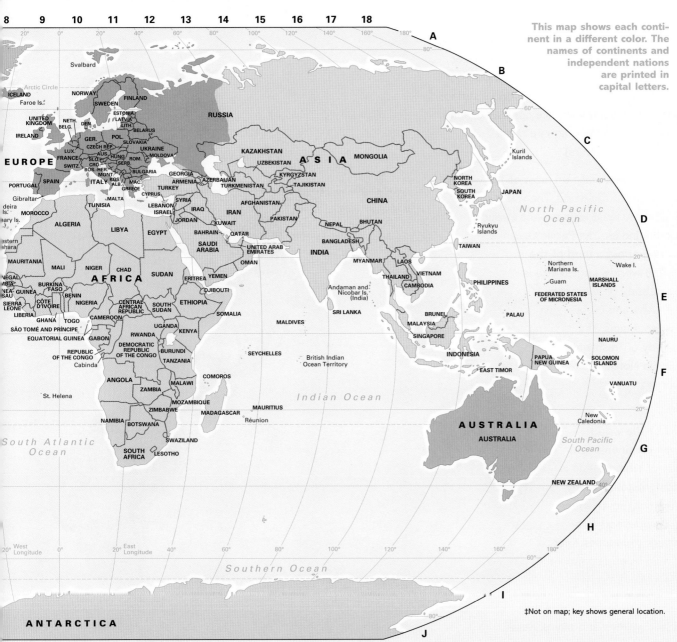

This map shows each continent in a different color. The names of continents and independent nations are printed in capital letters.

‡Not on map; key shows general location.

Name	Map key		Name	Map key		Name	Map key		Name	Map key		Name	Map key	
Guyana	E	7	Lebanon	D	11	Namibia	G	10	St. Vincent and the Grenadines	E	6	Taiwan	D	16
Haiti	E	6	Lesotho	G	11	Nauru	F	18	Samoa	F	1	Tajikistan	D	14
Honduras	E	5	Liberia	E	9	Nepal	D	14	San Marino	C	10‡	Tanzania	F	11
Hungary	C	10	Libya	D	10	Netherlands	C	10	São Tomé and Principe	E	10	Thailand	E	15
Iceland	B	9	Liechtenstein	C	10‡	New Zealand	G	18	Saudi Arabia	D	12	Togo	E	9
India	D	13	Lithuania	C	11	Nicaragua	E	5	Senegal	E	9	Tonga	F	1
Indonesia	F	16	Luxembourg	C	10	Niger	E	10	Serbia	C	10	Trinidad and Tobago	E	6
Iran	D	12	Macedonia	C	11	Nigeria	E	10	Seychelles	F	12	Tunisia	D	10
Iraq	D	12	Madagascar	F	12	Norway	B	10	Sierra Leone	E	9	Turkey	D	11
Ireland	C	9	Malawi	F	11	Oman	E	12	Singapore	E	15	Turkmenistan	D	13
Israel	D	11	Malaysia	E	15	Pakistan	D	13	Slovakia	C	11	Tuvalu	F	1
Italy	C	10	Maldives	E	13	Palau	E	16	Slovenia	C	11	Uganda	E	11
Jamaica	E	6	Mali	E	9	Panama	E	5	Solomon Islands	F	18	Ukraine	C	11
Japan	D	16	Malta	D	10	Papua New Guinea	F	17	Somalia	E	12	United Arab Emirates	D	12
Jordan	D	11	Marshall Islands	E	18	Paraguay	G	7	South Africa	G	11	United Kingdom	C	9
Kazakhstan	C	13	Mauritania	D	9	Peru	F	6	Sudan	E	11	United States	C	4
Kenya	E	11	Mauritius	G	12	Philippines	E	16	Sudan, South	E	11	Uruguay	G	7
Kiribati	F	1	Mexico	D	4	Poland	C	10	Suriname	E	7	Uzbekistan	D	14
Korea, North	C	16	Moldova	C	11	Portugal	D	9	Swaziland	F	11	Vanuatu	F	18
Korea, South	D	16	Monaco	C	10‡	Qatar	D	12	Sweden	B	10	Vatican City	C	10‡
Kosovo	C	11	Mongolia	C	15	Romania	C	11	Switzerland	C	10	Venezuela	E	6
Kuwait	D	12	Montenegro	C	10	Russia	C	13	Syria	D	11	Vietnam	E	15
Kyrgyzstan	C	13	Morocco	D	9	Rwanda	F	11				Yemen	E	12
Laos	E	11	Mozambique	F	11	St. Kitts and Nevis	E	6				Zambia	F	11
Latvia	C	11	Myanmar	D	14	St. Lucia	E	6				Zimbabwe	G	11

The surface area of the world totals about 196,900,000 square miles (510,000,000 square kilometers). Water covers about 139,700,000 square miles (362,000,000 square kilometers), or 71 percent of the world's surface. Only 29 percent of the world's surface consists of land, which covers about 57,200,000 square miles (148,000,000 square kilometers).

Oceans, lakes, and rivers make up most of the water that covers the surface of the world. The water surface consists chiefly of three large oceans—the Pacific, the Atlantic, and the Indian. The Pacific Ocean is the largest, covering about a third of the world's surface. The world's largest lake is the Caspian Sea, a body of salt water that lies between Asia and Europe east of the Caucasus Mountains. The world's largest body of fresh water is the Great Lakes in North America. The longest river in the world is the Nile in Africa.

The land area of the world consists of seven continents and many thousands of islands. Asia is the largest continent, followed by Africa, North America, South America, Antarctica, Europe, and Australia. Geographers sometimes refer to Europe and Asia as one continent called Eurasia.

The world's land surface includes mountains, plateaus, hills, valleys, and plains. Relatively few people live in mountainous areas or on high plateaus since they are generally too cold, rugged, or dry for comfortable living or for crop farming. The majority of the world's people live on plains or in hilly regions. Most plains and hilly regions have excellent soil and an abundant water supply. They are good regions for farming, manufacturing, and trade. Many areas unsuitable for farming have other valuable resources. Mountainous regions, for example, have plentiful minerals, and some desert areas, especially in the Middle East, have large deposits of petroleum.

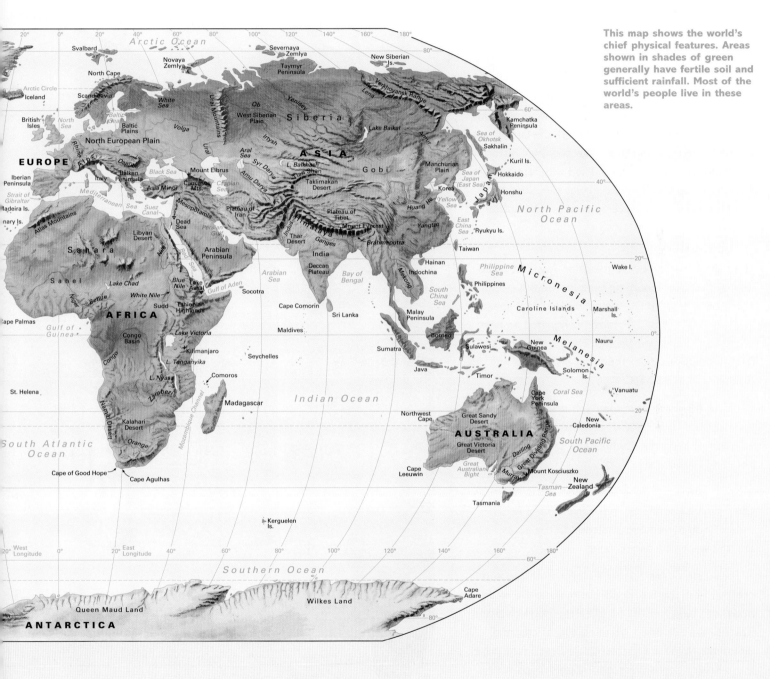

This map shows the world's chief physical features. Areas shown in shades of green generally have fertile soil and sufficient rainfall. Most of the world's people live in these areas.

EUROPE

ASIA

AFRICA

AUSTRALIA

ANTARCTICA

Arctic Ocean
North Pacific Ocean
South Pacific Ocean
Indian Ocean
South Atlantic Ocean
Southern Ocean

Svalbard
Severnaya Zemlya
New Siberian Is.
Novaya Zemlya
North Cape
Taymyr Peninsula
Scandinavia
Arctic Circle
Iceland
White Sea
Kamchatka Peninsula
British Isles
North Sea
Baltic Plains
Ob
West Siberian Plain
Siberia
Yenisey
Lena
Verkhoyansk Range
Sea of Okhotsk
Sakhalin
Volga
Ural Mountains
Baltic Sea
North European Plain
Irtysh
Lake Baikal
Amur
Kuril Is.
Rhine
Alps
Danube
Aral Sea
Ural
Hokkaido
Iberian Peninsula
Italy
Balkan Peninsula
Black Sea
Mount Elbrus
Syr Darya
Tien Shan
Gobi
Manchurian Plain
Sea of Japan (East Sea)
Honshu
Strait of Gibraltar
Asia Minor
Caucasus Mts.
Caspian Sea
Amu Darya
Taklimakan Desert
Korea
Yellow Sea
Madeira Is.
Mediterranean Sea
Suez Canal
Mesopotamia
Plateau of Iran
Plateau of Tibet
Huang He
East China Sea
Ryukyu Is.
Canary Is.
Atlas Mountains
Dead Sea
Persian Gulf
Himalaya
Mount Everest
Yangtze
Libyan Desert
Red Sea
Indus
Thar Desert
Ganges
Brahmaputra
Taiwan
Sahara
Arabian Peninsula
India
Hainan
Indochina
Wake I.
Sahel
Lake Chad
Gulf of Aden
Arabian Sea
Deccan Plateau
Bay of Bengal
Mekong
Philippine Sea
Micronesia
Niger
Blue Nile
Lake Assal
Socotra
Philippines
Benue
White Nile
Cape Comorin
South China Sea
Caroline Islands
Marshall Is.
Cape Palmas
Sudd
Ethiopian Highlands
Sri Lanka
Malay Peninsula
Nauru
Gulf of Guinea
Congo Basin
Lake Victoria
Maldives
Sumatra
Borneo
Sulawesi
New Guinea
Melanesia
Congo
Kilimanjaro
Seychelles
Java
Solomon Is.
St. Helena
L. Tanganyika
Comoros
Timor
Vanuatu
L. Nyasa
Coral Sea
Zambezi
Mozambique Channel
Madagascar
Indian Ocean
Northwest Cape
Great Sandy Desert
Cape York Peninsula
New Caledonia
Namib Desert
Kalahari Desert
Great Victoria Desert
Great Dividing Range
Orange
Great Australian Bight
Murray
Darling
Mount Kosciuszko
New Zealand
Cape of Good Hope
Cape Agulhas
Cape Leeuwin
Tasman Sea
Tasmania
Kerguelen Is.
West Longitude
East Longitude
Queen Maud Land
Wilkes Land
Cape Adare

Afghanistan is a completely landlocked nation, bordered by China on the far northeast, Pakistan on the east and south, Iran on the west, and Turkmenistan, Uzbekistan, and Tajikistan on the north. It is a country of rugged terrain and harsh, dry climate.

Because of its location, Afghanistan has long been a crossroads for migrating people and conquering armies. For centuries, it was the site of great empires and trading routes. All of these ancient peoples and empires left their mark on the culture of Afghanistan, creating a large number of ethnic groups who live in the country today.

Still, Afghanistan remains one of the least developed countries in the world. Most of its people still live in small, tribal villages and work the land using old-fashioned farming tools. The country's history has been long and troubled, filled with foreign invasions, political revolt, and civil war. Centuries of conflict have slowed the development of Afghanistan.

Ancient kingdoms

The region that is now Afghanistan was first inhabited by prehistoric hunting people 100,000 years ago. About 1500 B.C., Aryans invaded Afghanistan. In the mid-500's B.C., Persians conquered an area of Afghanistan called Bactria. The Persians ruled Bactria until Alexander the Great invaded Afghanistan in 330 B.C.

About 85 years later, the Bactrians revolted against Greek control and regained their land. Their kingdom lasted until the Kushans seized the country about 150 years later.

The Kushan Empire was founded in the A.D. 50's when Kujala Kadphises united five central Asian tribes. Soon the empire stretched to the Indus Valley and the western Ganges Valley.

The rulers of the Kushan Empire protected and profited from the Silk Road, the major trade route between China and the Middle East. Huge caravans traveled the Silk Road, bringing silk, spices, and ointments to India and the Middle East. From there, these luxury goods were loaded onto ships bound for the Roman Empire. In turn, the Romans sent back gold coins and Greek wine. New ideas and customs were also shared along these routes.

AFGHANISTAN

The Kushans were Buddhists. Twin statues of the Buddha, 174 feet (53 meters) high, once stood in central Afghanistan at Bamian. The statues were destroyed in 2001.

The arrival of Islam

In the late 600's, Arab conquerors swept into Afghanistan, defeating the Sassanians and the White Huns, who by then ruled the country. The Arabs brought with them the religion of Islam.

From the time of the Arab conquest until 1747, many armies fought over Afghanistan. Turkic-speaking people from eastern Persia and central Asia ruled from about 900 to 1200. The Mongols, led by Genghis Khan, conquered the country in the 1200's, as did the Timurids in the 1300's. The magnificent Blue Mosque at Mazar-e Sharif dates from the Timurid period.

From the mid-1500's to the early 1700's, Safavids from Persia and Mughals from India struggled for control. In 1747, Afghan tribes became united for the first time, under the leadership of Ahmad Shah Durrani. He established a monarchy that remained in power until 1973, when Afghanistan became a republic.

Even unified under a monarchy, Afghanistan still suffered from civil war between rival tribes. Foreign interference led to more wars on Afghan soil, as the United Kingdom and Russia fought for control of the country. The United Kingdom wanted to protect its empire in India, and Russia wanted an outlet to the Indian Ocean. In 1919, the United Kingdom ended its involvement, and Afghanistan became fully independent.

The Soviet Union sought to occupy Afghanistan in a war that lasted from 1979 to 1989. In the mid-1990's, a conservative Islamic group called the Taliban came to power. They allowed international terrorist organizations to create training camps in Afghanistan. Following devastating terrorist attacks against the United States in 2001, the United States and anti-Taliban forces within Afghanistan drove the Taliban from power. For the next decade, the United States led efforts to bring a lasting peace to Afghanistan under an elected government. However, in 2012, as U.S. and allied forces began to withdraw, attacks by the Taliban continued.

AFGHANISTAN TODAY

In recent years, Afghanistan has suffered bitter internal conflict. The current struggle has roots in 1973, when Prime Minister Muhammad Daoud led a revolt against King Muhammad Zahir, who was his relative. Military leaders took control of the government and established the Republic of Afghanistan. Daoud became president and prime minister.

Five years later, Daoud was killed when rival left wing military leaders and civilians revolted against the new government. This group had received a great deal of military and financial aid from the Soviet Union. When they took over the government, they established many Communist policies.

Civil war begins

Many Afghans disagreed with the Communist policies, claiming they violated Islamic teaching. They were also angered by the Soviet Union's influence on their government. These resistance fighters began to call themselves *mujahideen* (holy warriors). Fighting broke out between the mujahideen and the government. The Soviet Union sent troops into Afghanistan to help the government fight the mujahideen. The United States sent weapons to the mujahideen.

This conflict caused much death and destruction in Afghanistan. By the time the Soviets began to withdraw their troops in 1988, about 1-1/2 million Afghans had died.

An estimated 5 million Afghans lost their homes when the Soviets and the Afghan government bombed their villages. More than 3 million fled to refugee camps in Pakistan.

Throughout the struggle, the mujahideen were greatly outnumbered in air power, artillery, and tanks. However, their guerrilla tactics—hiding in the mountains and moving by night—confused their enemy. In the end, the Soviet and government troops controlled only main highways and urban centers.

When the Soviet occupation ended in 1989, most Afghans began to plan the rebuilding of their country's damaged economy. However, conflict between the mujahideen and the government continued.

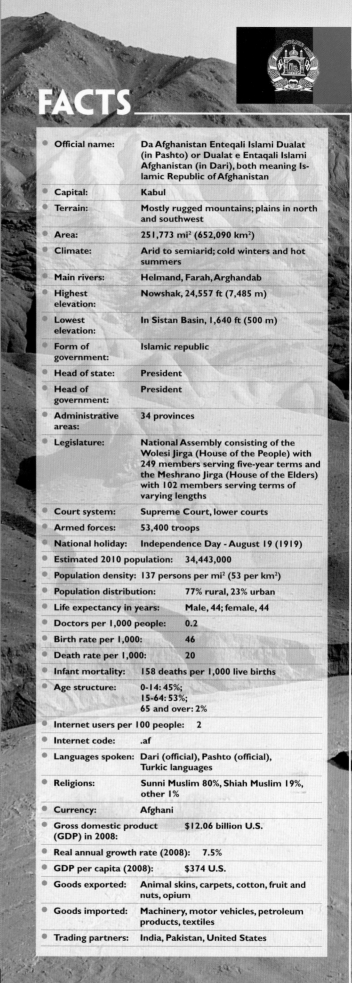

FACTS

Official name:	Da Afghanistan Enteqali Islami Dualat (in Pashto) or Dualat e Entaqali Islami Afghanistan (in Dari), both meaning Islamic Republic of Afghanistan
Capital:	Kabul
Terrain:	Mostly rugged mountains; plains in north and southwest
Area:	251,773 mi² (652,090 km²)
Climate:	Arid to semiarid; cold winters and hot summers
Main rivers:	Helmand, Farah, Arghandab
Highest elevation:	Nowshak, 24,557 ft (7,485 m)
Lowest elevation:	In Sistan Basin, 1,640 ft (500 m)
Form of government:	Islamic republic
Head of state:	President
Head of government:	President
Administrative areas:	34 provinces
Legislature:	National Assembly consisting of the Wolesi Jirga (House of the People) with 249 members serving five-year terms and the Meshrano Jirga (House of the Elders) with 102 members serving terms of varying lengths
Court system:	Supreme Court, lower courts
Armed forces:	53,400 troops
National holiday:	Independence Day - August 19 (1919)
Estimated 2010 population:	34,443,000
Population density:	137 persons per mi² (53 per km²)
Population distribution:	77% rural, 23% urban
Life expectancy in years:	Male, 44; female, 44
Doctors per 1,000 people:	0.2
Birth rate per 1,000:	46
Death rate per 1,000:	20
Infant mortality:	158 deaths per 1,000 live births
Age structure:	0-14: 45%; 15-64: 53%; 65 and over: 2%
Internet users per 100 people:	2
Internet code:	.af
Languages spoken:	Dari (official), Pashto (official), Turkic languages
Religions:	Sunni Muslim 80%, Shiah Muslim 19%, other 1%
Currency:	Afghani
Gross domestic product (GDP) in 2008:	$12.06 billion U.S.
Real annual growth rate (2008):	7.5%
GDP per capita (2008):	$374 U.S.
Goods exported:	Animal skins, carpets, cotton, fruit and nuts, opium
Goods imported:	Machinery, motor vehicles, petroleum products, textiles
Trading partners:	India, Pakistan, United States

Continued violence

In April 1992, the mujahideen overthrew the government. Several *factions* (small groups) within the mujahideen agreed to set up a transitional government. However, continued fighting among the mujahideen groups prevented the formation of a lasting, stable government.

In 1996, a military force of conservative Islamic youths, known as the Taliban (seekers of truth), captured Kabul and claimed power as the new national government. They set up a new Council of Ministers to rule the country. In 1997, Taliban authorities changed the official name of the country to the Islamic State of Afghanistan. But only three countries recognized the Taliban as a legal government—Pakistan, Saudi Arabia, and the United Arab Emirates.

In 1998, the United States accused the Taliban of harboring the Saudi millionaire Osama bin Laden, who was wanted in connection with terrorist attacks against two U.S. embassies in Africa. The United States launched missile strikes against suspected terrorist training camps in Afghanistan. The Taliban acknowledged that bin Laden was in Afghanistan under their protection. In 1999, the United Nations imposed trade sanctions against Afghanistan for refusing to surrender bin Laden.

In 2001, members of al-Qa`ida, the terrorist organization led by bin Laden, attacked the World Trade Center in New York City and the Pentagon Building near Washington, D.C. The attacks killed thousands of people. The United States demanded that the Taliban hand over bin Laden and shut down Qa`ida terrorist training camps in Afghanistan. The Taliban refused to do so, and the United States and its allies launched a military campaign against the Taliban. The campaign included air strikes in support of Afghan rebels who opposed the Taliban. This support enabled the rebels to drive the Taliban from power later in 2001.

An international peacekeeping force arrived in Kabul in late 2001 and early 2002. However, in the absence of a strong central government, warlords and tribal groups continued to fight for territory and power. Small groups of Taliban and Qa`ida forces also continued to battle U.S., Afghan, and allied troops.

Meanwhile, the United Nations helped the leaders of Afghanistan's main factions develop plans for a new constitution and a more democratic government. In 2002, a *loya jirga* (grand council of Afghan leaders) met in Kabul. It created a transitional government and chose Hamid Karzai, of the dominant Pashtun ethnic group, as the president. In January 2004, another loya jirga adopted a permanent constitution. Karzai was elected president under the new constitution in October 2004.

In 2006, the North Atlantic Treaty Organization (NATO) took over peacekeeping and security duties for Afghanistan from the U.S.-led alliance. Taliban and Qa`ida forces continued to fight NATO and the U.S.-led alliance.

Karzai remained in office after the 2009 presidential election, despite the fact that a United Nations investigation revealed numerous voting irregularities. In 2011, the United States began to withdraw its troops from Afghanistan, though insurgent attacks continued.

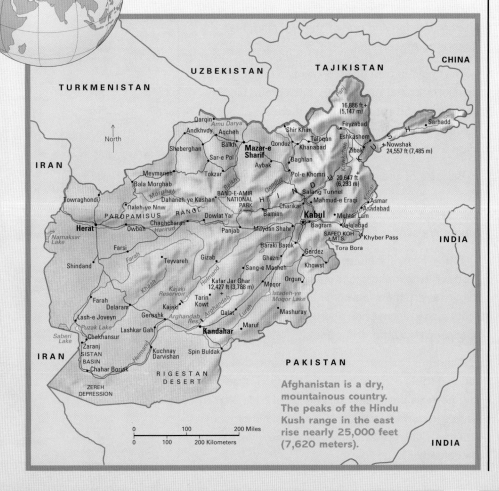

Afghanistan is a dry, mountainous country. The peaks of the Hindu Kush range in the east rise nearly 25,000 feet (7,620 meters).

PEOPLE

About 20 different ethnic groups, most with their own language and culture, call Afghanistan their home. They are united only by their devotion to Islam. Most Afghans are a blend of many early peoples who came to the country through migration or invasion. Many feel greater devotion and loyalty to their ethnic group than to their country.

Most Afghans live in rural areas. Since the Soviet invasion in 1979, several million Afghans have fled to Pakistan and Iran to escape wars. Thousands more Afghans have become refugees in their own country. About 4-1/2 million have returned to their homes since 2002.

The Pashtuns

The Pashtuns are the largest ethnic group in Afghanistan. They make up about half of the total population. Most Pashtuns live in eastern Afghanistan, in the mountainous areas near the Pakistan border. They speak Pashto, a language related to Persian and one of the two official languages of Afghanistan.

The Pashtuns are divided into two major groups, the Durranis and the Ghilzais. Each group has its own tribes and subtribes. In all, there are about 40 Pashtun tribes. Each tribe is led by a chief called a *khan*.

Because of their strong warrior tradition, the Pashtuns have a reputation for bravery in battle. During the 1800's and early 1900's, Pashtuns fought and won a series of wars with the British.

Today, many Pashtuns are farmers. They raise crops of wheat and other grains, fruits, nuts, and sugar cane. Others are nomadic herders, living in tents made of goat hair and tending their horses, sheep, goats, cattle, and camels.

Although Pashtuns are Sunni Muslims, their traditional tribal code of honor, rather than Islamic law, governs their daily life. This tribal code, known as *Pashtunwali*, teaches that personal honor is a Pashtun's most prized possession and must always be defended.

A rural tribesman, like most Afghan men, wears the traditional turban, which may be tied in a certain way to indicate an ethnic group.

Inhabitants of the city of Mazar-e Sharif gather in front of the Blue Mosque. Inside this blue-tiled mosque is the tomb of Ali, son-in-law of the prophet Muhammad. Built in the 1500's, it is an important holy site for Shiah Mus-lims and other Muslims. Shi`ah Muslims make up a small minority of the population. Most Afghans are Sunni Muslims.

The Tajiks

The Tajiks are the second largest ethnic group in Afghanistan, making up about 25 percent of the population. Like the Pashtuns, the Tajiks are Sunni Muslims. They live in central and northeastern Afghanistan. Their language is Dari, the country's other official language, which comes from *Farsi* (Persian).

The Tajiks do not live in tribal groups. Some are farmers and herders. Many live in the cities, where they work as craft workers, merchants, or administrators.

Other ethnic groups

The Uzbeks and the Turkomans are Turkic-speaking people who live in north and north-western Afghanistan. Many Uzbeks are farmers, and some are skilled craft workers widely known for their embroidered silk coats called *chapan*. The Turkomans still live in tribal groups.

Other ethnic communities include the Balochis, the Chahar Aimaks, and the Brahui. A group called the Hazaras live in the south-central mountain ranges and are said to be descended from the Mongols. Along with the Kizilbash, a smaller group, the Hazaras are Shiah Muslims.

High in the mountain valleys of the Hindu Kush in eastern Afghanistan live the Nuristanis. According to legend, the Nuristanis are descended from the Greek warriors who marched into Afghanistan under Alexander the Great in 330 B.C. Until the Nuristanis converted to Islam, they were known as *Kafirs* (infidels). After their conversion in the 19th century, they were given the name of Nuristanis, which means "people of the light."

A Kirghiz shepherd boy guards a flock of sheep and goats in a mountain valley. This small, semi-nomadic ethnic group roams the remote northeast.

A rug merchant in Herat relaxes with his stock. Herat is one of the oldest trading centers in Asia. It lies on a plain along the caravan route between Iran and India.

LAND AND ECONOMY

Afghanistan is a rugged, arid, mountainous country with dramatic scenery. The country also has many rivers, river basins, lakes, and desert areas.

Landscape and climate

Afghanistan can be divided into three land regions: (1) the Northern Plains, (2) the Central Highlands, and (3) the Southwestern Lowlands.

The Northern Plains extend across northern Afghanistan. They consist of mountain plateaus and rolling hills.

Although the soil in the region is fertile, the land can be cultivated only in river valleys and mountain areas, where water is available. Afghan farmers have built large irrigation systems along the Harirud, Helmand, Qonduz, and other rivers to provide water for their crops.

Seminomadic people also live in the Northern Plains. They wander on its vast grasslands with their flocks of sheep and goats.

Summers in the Northern Plains are hot and dry, with average temperatures of about 90° F (32° C). Winters are cold and dry, with average temperatures of about 38° F (3° C).

The Central Highlands consist of the Hindu Kush mountain range and its branches. This region covers about two-thirds of Afghanistan. Most Afghans live in the high, narrow valleys of the Hindu Kush.

Winters in the Central Highlands are cold, with average temperatures of about 25° F (– 4° C) in January. Summers are mild, with average temperatures of 75° F (24° C).

The Southwestern Lowlands are mainly desert or semi-desert. The Helmand River crosses the region, flowing from the Hindu Kush to the Sistan Basin on the Iranian border. Barley, corn, fruits, and wheat are grown in the Helmand Valley.

Temperatures in the Southwestern Lowlands average about 35° F (2° C) in January and about 85° F (29° C) in July. The Sistan Basin suffers from a crop-destroying summer wind called the "wind of 120 days."

Opium poppies are a profitable crop for Afghan farmers. The poppies are used to make the illegal drug heroin.

Cave dwellings are carved into high cliffs that overlook the fields of the fertile Bamian valley.

The economy

Even though little of Afghanistan's land is suitable for farming, most of the country's people work the land for a living. To water their fields, Afghan farmers depend on river irrigation systems and underground springs and streams called *qanats*.

Wheat is Afghanistan's chief crop. Other crops include barley, corn, cotton, fruits, nuts, rice, and vegetables. The chief livestock products are dairy items, mutton, wool, animal hides, and the skins of karakul sheep.

Afghanistan's rich mineral resources are largely undeveloped. Since natural gas was discovered in the northern Sheberghan area during the 1960's, it has become an important part of the economy. Afghan mines produce some of the world's finest *lapis lazuli,* a valuable blue gemstone. Afghanistan also produces some coal, copper, gold, and salt.

Violent conflict in Afghanistan has severely damaged the country's economy. Because so many Afghans fled the country, much farmland has been left uncultivated. In addition, irrigation systems have been neglected or damaged by the fighting.

Painted Afghan trucks, above, carry goods through crowded streets as well as through mountain passes.

Carpet weaving is part of a large handicraft industry. Rugs from Afghanistan are known for having rich, deep colors such as red and navy.

Farming and raising livestock are how most Afghans make their living. The people who live off the land make their homes in the high mountain valleys and on the broad, rolling grasslands that cover much of the country.

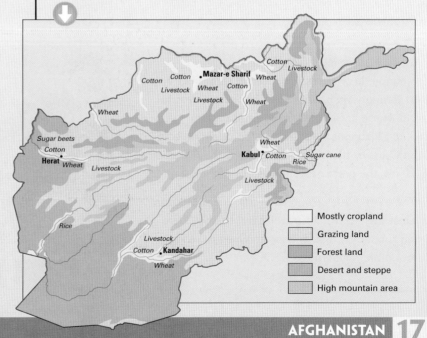

Legend:
- Mostly cropland
- Grazing land
- Forest land
- Desert and steppe
- High mountain area

THE KHYBER PASS

The Khyber Pass is a passage about 30 miles (48 kilometers) long through the Hindu Kush mountain range. It connects the northern frontier of Pakistan with Afghanistan. At its narrowest point, the pass is only 50 feet (15 meters) wide. On the north side of the Khyber Pass rise the towering, snow-covered mountains of the Hindu Kush. The name *kush*, which means *death*, may have been given to this range because of its dangerous passes.

The Khyber Pass is one of the most famous mountain passes in the world. It is the best land route between India and Pakistan and has had a long and often violent history. Conquering armies have used the pass as an entry point for their invasions. It has also been a major trade route for centuries.

Invading armies

During the 320's B.C., Alexander the Great and his army probably marched through the Khyber Pass when they invaded the upper Indus Valley, in what is now Pakistan. Many other invaders, including the central Asian leaders Timur in the A.D. 1300's and Babur in the 1500's, also crossed the Khyber Pass on their way into South Asia. Centuries later, India became part of the British Empire, and British troops defended the pass. During the 1800's, the pass played an important role in the struggle between the United Kingdom and Russia for control of Afghanistan.

Caravans from China

For hundreds of years, great camel caravans traveled through the Khyber Pass, bringing goods to trade. These ancient merchants and traders brought luxurious silks and fine porcelain objects from China to the Middle East. Often, they stopped at Herat, the great oasis in western Afghanistan.

The traders traveled in caravans as a protection against the hazards of travel. Even so, they were often robbed by local tribesmen when traveling through the Khyber Pass.

The Khyber Pass is one of the most important passes between Afghanistan and Pakistan. It is known as the gateway to India. Ancient Persians and Greeks, Tatars, and Moguls passed through the Khyber to reach India.

A crowded bus travels through the Khyber Pass over a highway that links Kabul in Afghanistan with Peshawar in Pakistan. Another highway is reserved for caravans.

The Khyber Pass today

Today, two highways thread their way through the Khyber Pass—one for motor traffic, and one for the traditional caravans. A railway line also travels to the head of the pass. Millions of Afghan refugees have fled to Pakistan through the Khyber Pass to escape from the warfare in Afghanistan during the 1980's, 1990's, and early 2000's.

Villages lie on each side of the Khyber Pass. The people of the Khyber region are mainly Pashtuns. They live in both Afghanistan and Pakistan.

The Pashtuns consist of about 40 tribes divided into groups of related families. Although the tribes unite to fight invaders, they often feud with one another. Some of the feuds have raged for centuries. The Pashtuns are known for their traditional gunmaking skills.

A Pashtun tribesman is ready for battle, whether defending his personal honor or serving as a resistance fighter.

The Khyber Pass links Afghanistan and Pakistan. The mountain pass is about 33 miles (53 kilometers) long and cuts through the Safed Koh mountains in the Hindu Kush range. Over the centuries, many travelers, traders, and invaders have followed this route.

ALBANIA

Albania is a small, mountainous nation on the Balkan Peninsula and one of the least developed countries in Europe. It became a Communist country in 1944. Until the breakup of the Communist Party's control in 1990, Albania was one of the most politically isolated countries in the world.

Early history

In ancient times, much of the territory that is now Albania was settled by the Illyrians. In 167 B.C., the Romans conquered the Illyrians and spread their civilization throughout the region. In A.D. 395, when the Roman Empire was divided, much of Albania became part of the East Roman (Byzantine) Empire. Between the 300's and the 1000's, Albania was invaded by Goths, Bulgarians, Slavs, and Normans. In the 1300's, much of Albania became part of the Serbian Empire.

Between 1468 and 1478, the Ottomans conquered Albania, and it remained part of the Ottoman Empire for more than 400 years. In 1912, Albania gained independence during the First Balkan War. The next year, European powers set the country's boundaries and made it a self-governing principality.

During World War I (1914–1918), the Austro-Hungarians, Italians, Serbs, and French all occupied Albania. In 1925, Ahmed Beg Zogu seized power, formed a republic, and became the country's first president. Three years later, Zogu declared himself King Zog I and ruled as a dictator until 1939.

Shortly before World War II (1939–1945), Italy invaded Albania, and Albania became part of the Italian Empire. After Italy surrendered to the Allies in 1943, the Germans occupied Albania until they were driven out the following year.

The rise of Communism

During World War II, Albania found a new national hero in Enver Hoxha, leader of the National Liberation Front (NLF). The NLF, a Communist organization, was one of the Albanian resistance

FACTS

Official name:	Republika e Shqiperise (Republic of Albania)
Capital:	Tiranë
Terrain:	Mostly mountains and hills; small plains along coast
Area:	11,100 mi^2 (28,748 km^2)
Climate:	Mild temperate; cool, cloudy, wet winters; hot, clear, dry summers; interior is cooler and wetter
Main rivers:	Bunë, Drin, Mat, Shkumbin, Vjosë
Highest elevation:	Mount Korab, 9,068 ft (2,764 m)
Lowest elevation:	Adriatic Sea, sea level
Form of government:	Democracy
Head of state:	President
Head of government:	Prime minister
Administrative areas:	12 qarqe (counties)
Legislature:	Kuvendi Popullor (People's Assembly) with 140 members serving four-year terms
Court system:	Constitutional Court, Supreme Court
Armed forces:	14,300 troops
National holiday:	Independence Day - November 28 (1912)
Estimated 2010 population:	3,245,000
Population density:	292 persons per mi^2 (113 per km^2)
Population distribution:	54% rural, 46% urban
Life expectancy in years:	Male, 73; female, 80
Doctors per 1,000 people:	1.2
Birth rate per 1,000:	15
Death rate per 1,000:	6
Infant mortality:	13 deaths per 1,000 live births
Age structure:	0-14: 25%; 15-64: 66%; 65 and over: 9%
Internet users per 100 people:	15
Internet code:	.al
Languages spoken:	Albanian (official), Greek, Vlach, Romani, Slavic dialects
Religions:	Muslim 70%, Albanian Orthodox 20%, Roman Catholic 10%
Currency:	Lek
Gross domestic product (GDP) in 2008:	$12.93 billion U.S.
Real annual growth rate (2008):	6.1%
GDP per capita (2008):	$4,056 U.S.
Goods exported:	Chromite and other minerals, footwear, fruits and vegetables, textiles
Goods imported:	Chemicals, farm and industrial machinery, food, textiles
Trading partners:	Germany, Greece, Italy, Turkey

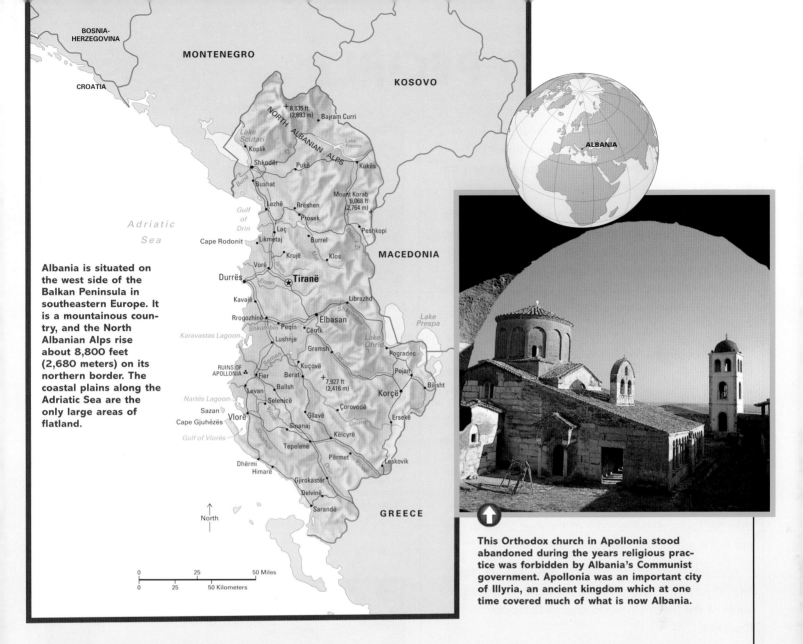

Albania is situated on the west side of the Balkan Peninsula in southeastern Europe. It is a mountainous country, and the North Albanian Alps rise about 8,800 feet (2,680 meters) on its northern border. The coastal plains along the Adriatic Sea are the only large areas of flatland.

This Orthodox church in Apollonia stood abandoned during the years religious practice was forbidden by Albania's Communist government. Apollonia was an important city of Illyria, an ancient kingdom which at one time covered much of what is now Albania.

movements that fought against the Italians and the Germans. In 1944, the Germans were driven from Albania, and Hoxha set up a Communist government and began ruling as first secretary of the Communist Party. He ruled Albania for more than 40 years, until his death in 1985.

Ramiz Alia succeeded Hoxha as first secretary of Albania's Communist Party. His government introduced some social and economic reforms. However, Albanians felt that the reforms did not go far enough. Their protests in 1990 led the Communists to allow new political parties to form.

In the multiparty elections of March 1991, the Communists won a majority, but protests continued. By June, the Communist prime minister and his Cabinet were forced to resign. A temporary government was formed, and new elections were held in March 1992. The Democratic Party, the leading opposition group, won a majority of seats in the National Assembly.

In late 1996, the collapse of fraudulent nationwide pyramid schemes triggered civil unrest that turned to armed rebellion in 1997. Thousands of Albanians fled to Greece and Italy. In April, the United Nations (UN) sent an international force to Albania to help restore order. Elections held in June and July brought the Socialist Party into power, and the UN left in August. In 1998, a new constitution was adopted. The Democratic Party won the most seats in the parliamentary elections held in 2005 and 2009.

LAND AND PEOPLE

Albania is bordered on the west by the Adriatic Sea, on the northwest by Montenegro, on the northeast by Serbia, on the east by Macedonia, and on the southeast and the south by Greece. Most of the land is mountainous.

In the northern part of the country, the North Albanian Alps gradually give way to thick forests of oak, elm, and conifers in the central uplands. The central uplands flatten out to the west and form the Adriatic coastal plain. The mountains of the south extend down to the Albanian Riviera, where their sheer cliffs plunge into the Adriatic Sea.

Albania's coastal region has a Mediterranean climate, with warm, dry summers and mild, wet winters. Rain falls mostly in the winter, but the distribution is uneven, with the northern mountains receiving the most rain.

People

Albanians are divided into two major groups—the Ghegs and the Tosks. Most of the Ghegs live north of the Shkumbin River. They speak a different Albanian dialect than the Tosks, who live south of the river. Some Greeks live in the regions that adjoin Greece.

Approximately 70 percent of all Albanians are Muslims. Most of the rest, including the country's Greeks and some of the Tosks, belong to Eastern Orthodox Churches.

About half of the Albanian people live on farms or in rural villages. Some people along the Adriatic coast earn their living by fishing. Most Albanian towns and cities are small.

Living standards in Albania are extremely low compared with those of other European countries. The incomes of most Albanians are small, but health care, social services, and education are free. Bread, vegetables, and such dairy products as cheese and milk make up the daily diet of most of the people.

Buses and trains are common means of transportation in Albania. Under the country's former Communist government, it was illegal for private individuals to own an automobile. Private ownership of an automobile only became legal in 1991.

A farmer ploughs his land in the Vjosa Valley in southeastern Albania. The Vjosa River, one of Albania's main waterways, runs through the valley. Major crops in Albania include corn, grapes, olives, potatos, sugar beets, vegetables, and wheat.

Young people pause along a path in the northern Albanian countryside to chat with the village elders. Since 1945, Albania's education system has improved, and illiteracy has ben greatly reduced.

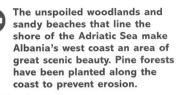

The unspoiled woodlands and sandy beaches that line the shore of the Adriatic Sea make Albania's west coast an area of great scenic beauty. Pine forests have been planted along the coast to prevent erosion.

Pedestrians crowd the sidewalks and streets of Tiranë, Albania's capital and largest city. Most Albanians use buses, bicycles, and trains to get around because few people can afford the luxury of owning an automobile.

Agriculture and industry

Agriculture is an important sector of Albania's economy, and it employs more Albanians than any other economic activity. The most important farming regions are located on the Adriatic coastal plain. The valleys in the interior of the country, particularly Korcë in the southeast region, are also used for agricultural purposes.

Albania's chief crops are corn, grapes, olives, potatoes, sugar beets, vegetables, and wheat. Farmers also raise livestock, including cattle, sheep, goats, pigs, and chickens. Under the Communist government, all agriculture in Albania was organized into state or collective farms. The government tried to increase agricultural production by introducing modern equipment and chemical fertilizers. In 1991, the new government took steps toward breaking up the socialist agricultural system and establishing private ownership of the land. Today, most farmland is privately owned.

Albania has relatively few factories. Its factory products include cement, food products, and textiles. Many of these light industries are centered in Tiranë, Albania's capital and largest city.

Albania is rich in mineral deposits, and mining is the leading industrial activity. Albania is an important producer and exporter of chromite, and *lignite* (brown coal) is mined in central and southern Albania. Albania also produces copper, natural gas, and petroleum.

Albania's exports include chromite and other minerals, footwear, fruits and vegetables, and textiles. Chemicals, farm and industrial machinery, food, and textiles are among its chief imports. Albania imports much more than it exports. Italy is its main trading partner. *Remittances* (money sent home) from Albanians living in Greece and Italy are also important to the economy.

Traditionally, Algeria was part of the Maghreb—the North African lands that also included Morocco, Tunisia, and part of Libya. The name *Maghreb,* an Arabic word meaning *the place of the sunset—the west,* reflects Algeria's Arab history.

People have lived in what is now Algeria for at least 40,000 years. By about 3000 B.C., nomadic Berbers had begun migrating into the region—probably from Europe or Asia. During the 1100's B.C., the Phoenicians sailed from the eastern Mediterranean and established trading posts on the Algerian coast.

About 200 B.C., the Romans helped a Berber chieftain named Massinissa form a kingdom called Numidia in northern Algeria, where he ruled as king. The land was part of the Roman Empire from 46 B.C. until the Vandals, a barbarian tribe from northern Europe, took control of the country in the A.D. 400's. Later, Byzantines ruled the area.

During the A.D. 600's, Arabs invaded the Maghreb. The Arabs brought their religion, Islam, to the region. In time, the Berbers became *Muslims* (followers of Islam), and the Arabic culture and language spread throughout the region.

Spanish Christians captured coastal towns in Algeria during the early 1500's, but in 1518, a Turkish sea captain named Barbarossa took control of Algiers and drove the Spaniards out of most of Algeria. Barbarossa then joined the areas under his control with the Ottoman Empire, an Islamic empire based in Turkey.

Algeria remained part of the Ottoman Empire until the early 1800's. During that period, Algerian pirates called *corsairs* provided Algeria with its main source of income by attacking and looting ships in the Mediterranean Sea.

In 1830, France gained control of northern Algeria. European settlers, given French citizenship and large sections of Algerian land, soon controlled Algeria's economy and government. Many Algerians rebelled against the French, but in 1847, the French defeated the rebel forces under Abd al-Qadir, a Muslim religious leader. By 1914, France controlled all of Algeria.

As part of France, Algeria fought on the side of the Allies during World War I (1914–1918). During World War II (1939–1945), Algeria itself became a battleground. In 1940, France surrendered to the invading

ALGERIA

German forces of Adolf Hitler, and cooperative French officials set up a government under German control in Vichy, France. The Vichy government ruled Algeria until 1942, when Allied forces invaded Algeria and Morocco. After the war, control of Algeria was returned to France. When France refused to give Algerians a greater voice in their government, many Algerians began to demand independence.

In 1954, a revolution was launched by the Algerian Front de Libération Nationale (FLN), or National Liberation Front. A long and bloody war followed. On July 3, 1962, France finally granted Algeria its independence.

Since independence, Algeria's population has grown rapidly, and many poor rural people have moved to cities to seek factory work. However, many Berbers live in mountain villages and depend on farming or herding to make their living. Industry has not grown fast enough to eliminate such problems as poverty, housing shortages, and widespread unemployment.

Algeria's past continues to influence many aspects of life today. The majority of Algerians speak Arabic. Many also speak French, even though the role of French culture in Algerian society decreased greatly after independence. About a fifth of the people speak dialects of Tamazight, the Berber language.

Algerian painters and writers were strongly influenced by French culture during the period when Algeria belonged to France. Since then, artists have increasingly drawn upon their religious and ethnic roots. Jewelry, pottery, rugs, and other handicrafts often feature distinctive Islamic designs and are crafted with traditional techniques.

Islam is Algeria's official religion. The government pays for the maintenance of mosques and for the training of mosque officials. Not all Algerians agree, however, on the role that Islam should play in the country's political and social life. In 2011, antigovernment protests erupted throughout Algeria. The protesters called for greater political freedom.

ALGERIA TODAY

Algeria is the largest country in Africa. Algiers, the nation's capital and largest city, lies on the Mediterranean coast. South of the narrow coastal region, on the other side of the Atlas Mountains, lies the vast desert land of the Sahara.

During and after Algeria's violent struggle for independence, most of the European settlers left the country. Algeria became a democratic republic in 1962. A rebel leader, Ahmed Ben Bella, became Algeria's first president. Ben Bella declared Algeria a socialist state.

In 1965, Ben Bella was overthrown by Houari Boumedienne, an army commander. Boumedienne launched a program of rapid economic development, using the money from oil and natural gas production. An Algerian constitution adopted in 1976 proclaimed the FLN, the group that launched the fight for independence, the country's only political party.

After Boumedienne died in 1978, Defense Minister Chadli Bendjedid was elected president. He was re-elected in 1984 and 1988.

In 1989, the constitution was revised to allow other parties to run candidates. However, the main opposition party, the Islamic Salvation Front (FIS), claimed that changes in the electoral law still favored the FLN. During elections for the National People's Assembly in late 1991, it appeared that the FIS would win a majority of seats. In 1992, the military seized power to prevent an FIS election victory and banned the FIS. A military-dominated committee governed Algeria until 1994, when Liamine Zeroual was chosen to be president. Zeroual won a multiparty election for president in 1995.

In 1996, the Algerian people approved a revised constitution that bans political parties based on religion, sex, language, or regional differences. Algeria held multiparty parliamentary elections in 1997. A party called the National Democratic Rally won the most seats. In presidential elections held in 1999, independent candidate Abdelaziz Bouteflika was elected, though some observers made charges of vote fraud. Bouteflika was reelected in 2004 and 2009.

FACTS

Official name:	Al Jumhuriyah al-Jaz'iriyah ad Dimuqratiyah wa ash-sha'biyah (People's Democratic Republic of Algeria)
Capital:	Algiers
Terrain:	Mostly high plateau and desert; some mountains; narrow, discontinuous coastal plain
Area:	919,595 mi² (2,381,741 km²)
Climate:	Arid to semiarid; mild, wet winters with hot, dry summers along coast; drier with cold winters and hot summers on high plateau
Main river:	Chelif
Highest elevation:	Mount Tahat, 9,573 ft (2,918 m)
Lowest elevation:	Chott Melrhir, 102 ft (31 m) below sea level
Form of government:	Republic
Head of state:	President
Head of government:	Prime minister
Administrative areas:	48 wilayas (provinces)
Legislature:	Parliament consisting of the Al-Majlis Ech-Chaabi Al-Watani (National People's Assembly) with 389 members serving five-year terms and the Council of Nations with 144 members serving six-year terms
Court system:	Cour Supreme (Supreme Court)
Armed forces:	147,000 troops
National holiday:	Revolution Day - November 1 (1954)
Estimated 2010 population:	35,415,000
Population density:	39 persons per mi² (15 per km²)
Population distribution:	64% urban, 36% rural
Life expectancy in years:	Male, 72; female, 75
Doctors per 1,000 people:	1.1
Birth rate per 1,000:	21
Death rate per 1,000:	5
Infant mortality:	28 deaths per 1,000 live births
Age structure:	0-14: 28%; 15-64: 67%; 65 and over: 5%
Internet users per 100 people:	10
Internet code:	.dz
Languages spoken:	Arabic (official), French, Berber dialects
Religions:	Sunni Muslim (state religion) 99%, Christian and Jewish 1%
Currency:	Algerian dinar
Gross domestic product (GDP) in 2008:	$168.28 billion U.S.
Real annual growth rate (2008):	3.0%
GDP per capita (2008):	$4,898 U.S.
Goods exported:	Natural gas, petroleum and petroleum products
Goods imported:	Food, machinery, raw materials
Trading partners:	France, Germany, Italy, United States

In 1999, the Islamic Salvation Army, the armed branch of the FIS, announced it was ending its fight against the government. From 1992 to 2006, more than 150,000 people died in the fighting. Although most of the fighting stopped, some Muslim extremists continued to attack Algerian security forces and civilians.

In 2001, violent clashes broke out between security forces and Berber protesters in northern Algeria. The protesters demanded greater political and cultural recognition for Berbers. The government agreed to some Berber demands. In 2002, for example, the government made Tamazight a national language.

In 2011, antigovernment protests erupted in Algiers and other cities. The protesters called for greater political freedom. The unrest followed similar events in Tunisia, Egypt, and elsewhere in the region.

A street vendor sells French baguettes (long, crusty loaves of bread) in the dusty square of an Algerian town. Algeria, once a French colony, retains such French influences, in spite of the government's efforts to remove them. In 1970, a "cultural revolution" was launched to help Algerians recover their national identity.

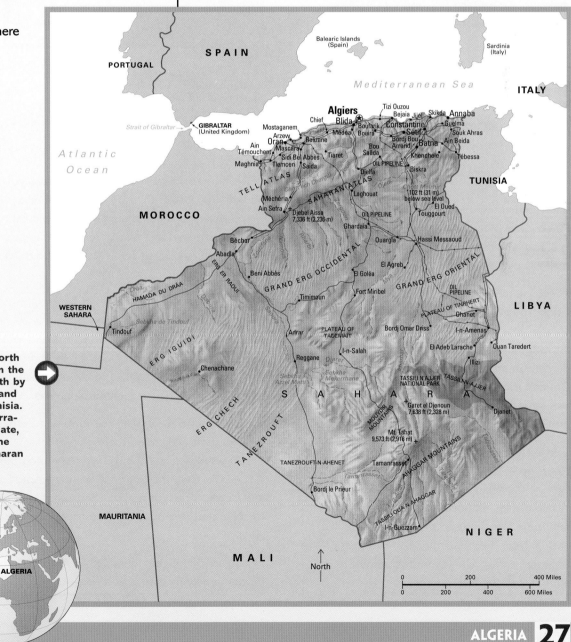

Algeria is bordered on the north by the Mediterranean Sea; on the west by Morocco; on the south by Mauritania, Mali, and Niger; and on the east by Libya and Tunisia. The country's narrow Mediterranean region has a warm climate, rich farmland, and most of the population. South of the Saharan Atlas Mountains, the great Sahara covers more than 80 percent of Algeria.

LAND AND PEOPLE

The northern coastal area of Algeria, called the Tell, stretches along the Mediterranean Sea. This region has the mild temperatures and moderate rainfall typical of a Mediterranean climate. The Tell extends only about 80 to 200 miles (130 to 320 kilometers) south of the coast, but more than 90 percent of the Algerian people live on it.

The Tell is Algeria's heartland. The word *tell* means *hill* in Arabic, and the region is aptly named for its gently rolling hills and coastal plains. Much of Algeria's best farmland lies in the western and central Tell. Rugged mountains make up the eastern Tell, and the Tell Atlas Mountains rise along the region's southern edge.

South of this range lie the High Plateaus, which are cooler and drier than the Tell at about 1,300 to 4,300 feet (400 to 1,300 meters) above sea level. The plateaus are home to about 7 percent of the Algerian people.

The High Plateaus end with the Saharan Atlas Mountains. To the south lies the barren, sun-baked Sahara, the largest desert on Earth. Fewer than 3 percent of the Algerian people live in this desolate region, which covers more than 80 percent of the country's land.

The vast Sahara actually has a variety of landscapes. Sand dunes cover much of the northern Sahara. Two huge seas of sand, called *ergs,* dominate this area: the Grand Erg Occidental (Great Western Erg) and the Grand Erg Oriental (Great Eastern Erg). Other parts of the Sahara include vast stretches of bare rock, boulders, and gravelly stone.

In southeast Algeria, the Ahaggar Mountains tower up to 9,573 feet (2,918 meters). Northeast of the Ahaggar, in a highland area called Tassili-n-Ajjer, needlelike rocks point to the sky, and huge, petrified sand formations rise like castles from the desert floor. To the west of the Ahaggar, an almost lifeless pebble desert stretches to the Mali border.

⬆

A massive rock formation looms above the barren landscape near the Ahaggar Mountains in southeast Algeria. The Ahaggar, together with the highland region Tassili-n-Ajjer, help make the Saharan region of Algeria a land of great scenic diversity.

Daytime temperatures in the Sahara can soar above 120° F (49° C). During the summer, a hot, dusty wind called the *sirocco* blows north across the desert and blasts the High Plateaus. The sirocco reaches as far as the Tell about 20 days of the year.

The Tell's pleasant climate and fertile farmland have drawn most of Algeria's people to this area. More than 1.5 million people live in Algiers alone. Many rural Algerians have moved to the cities seeking work, and about two-thirds of the population now live in urban areas.

Most Algerians are of mixed Arab and Berber descent. Berbers settled the region at least 5,000 years ago, and Arabs invaded during the A.D. 600's. Over the centuries, so many Arabs and Berbers intermarried that today it is difficult to separate the groups by ancestry or appearance.

Instead, people are identified as Arab or Berber largely by their way of life—by the

language they speak and the customs they follow. About a fifth of Algeria's people speak Berber languages and follow Berber customs, so they are considered Berbers.

When the Arabs invaded Algeria, they brought their religion, Islam, with them. About 99 percent of Algerians today are Muslims. The teachings of Islam govern family relationships and other aspects of daily life.

Following Algerian Islamic tradition, men and women have lived vastly different lives. For example, women usually wear veils in public because it is considered improper for a woman's face to be seen by a man who is not related to her. As Algeria wrestles with Western ideas, some younger and more educated women resist such practices. Some city people follow other Western customs as well, wearing Western-style clothing and eating Western foods, for example.

An Algerian farmer shelters a young goat in the folds of his burnoose—a heavy, hooded cloak. More than half of all Algerians live in rural areas, mainly raising livestock or farming small plots of land for a living.

Heavily veiled women, with all but their eyes hidden by voluminous, white garments called haiks, walk through the streets of Beni Isguen, a small town in the north central Algerian desert. Many urban women in Algeria do not veil their faces.

Nomadic herders live in tents made of animal skins or mats thrown over a framework of wooden poles. This kind of shelter allows them to move easily in search of water and grazing land for their animals. Only a small percentage of the Algerian people live in the Sahara, and many of them have abandoned their nomadic lifestyles to settle in oases.

A shallow stream provides a valuable and scarce resource in Algeria—water. In the desert, dry streambeds called wadis fill with water for a short time after the infrequent rains.

ECONOMY

After Algeria won its independence in 1962, many French and other European settlers fled the country. These were the people who had been running Algeria's most modern farms and factories. Algerians then formed a *socialist* government—one that controlled the means of producing goods, such as the farms and factories. The government worked to develop industry, especially the petroleum industry. It used the money from industry to develop modern methods of agriculture and to produce consumer goods. Algeria became a leader among developing nations.

Today, Algeria is changing from an economy based on government control to more private ownership. The country's economy still is based largely on income from natural gas and petroleum production. The government controls the nation's natural gas and petroleum industry.

Mining

Algeria's huge deposits of petroleum and natural gas have been a major factor in its development. The northeastern part of the Sahara in Algeria has especially rich oil fields. Natural gas, petroleum, and petroleum products account for almost all the value of Algeria's exports.

Algeria belongs to The Organization of Petroleum Exporting Countries (OPEC), an association of countries whose economies depend heavily on oil exports.

Mining employs only about 2 percent of Algeria's workers, but it accounts for much of the country's economic production. In addition to natural gas and petroleum, minerals mined in Algeria include iron ore, lead, mercury, phosphate rock, and zinc.

Service, manufacturing, and construction

Service industries account for about one-third of the total value of Algeria's economic production. They employ about half of the nation's workers—

Camels wait patiently to be loaded with bricks of salt and bundles of grain. Camel caravans still carry goods and people across the great Sahara, though today more modern methods of transportation, such as jeeps and planes, are also used.

Oil workers rest from their labors in the shade of a tanker truck. The production and refining of petroleum and natural gas are the mainstay of the nation's economy.

many more people than mining employs. Service workers include people with jobs in banks, government agencies, hospitals, insurance companies, and schools.

Manufacturing and construction account for much of Algeria's economic production and employ about one-fourth of its workers. The nation's chief manufactured products include construction materials, iron and steel, liquid natural gas, motor vehicles, and refined petroleum products.

While most small factories are privately owned, the government has poured money into building factories and controls key manufacturing industries. Almost all factories are located in the Mediterranean coastal area in such cities as Algiers, Annaba, Arzew, Constantine, and Skikda. However, there are still not enough manufacturing jobs. As a result, Algeria has a high rate of unemployment. Many Algerians work in foreign countries.

Hassi Messaoud in eastern Algeria is the first and largest oil field in Algeria. A giant refinery was built over the field in the eastern part of the country in 1979. Pipelines carry oil from the field to cities in northern Algeria.

An abundant date harvest awaits transportation from the oasis of Touggourt in Algeria's northeastern Sahara. Date palms in oases are watered by underground springs.

Agriculture

Agriculture provides a living for about one-fifth of Algeria's workers. Most farmers own small plots of land that produce only enough to feed their own families. Some farmers work on large government farms. Algeria has little arable land, however, and must import much of its food.

The western and central Tell have the nation's best farmland. Grains, especially wheat and barley, are Algeria's chief crops. Farmers also produce citrus fruits, dates, grapes, olives, potatoes, and tomatoes. Many people in the High Plateaus herd cattle, sheep, and goats, providing dairy products and meat.

Transportation and communication

Algeria has tens of thousands of miles of roads and thousands of miles of railroad track, and nearly all of it lies north of the Sahara. Camel caravans still wind across the desert, transporting goods and people as they have done for centuries. But today, planes, jeeps, and trucks are also used.

The government controls some of the country's daily newspapers, but others are privately owned. The government operates all the radio and television stations.

AMERICAN SAMOA

A territory of the United States, American Samoa lies south of the equator, about 2,300 miles (3,700 kilometers) southwest of Hawaii. The seven islands that make up American Samoa have a total area of 76 square miles (197 square kilometers).

Six of the territory's seven islands are divided among three groups—Tutuila and Aunuu; Ofu, Olosega, and Tau; and Rose. These islands are in the Samoan chain. The seventh, Swains Island, lies 200 miles (310 kilometers) north. Tutuila, the largest and most important island, lies at the western end of American Samoa. Pago Pago (pronounced *PAHNG oh PAHNG oh*), American Samoa's capital and only urban center, is on Tutuila. Pago Pago has one of the best and most beautiful harbors in the South Pacific.

Rose and Swains islands are coral islands, while the others are the remains of extinct volcanoes. American Samoa has a wet, tropical climate. Only a third of the territory's land can be cultivated. Most of the land is mountainous, with some fertile soil in the valleys. The islands have few natural resources.

People

Almost all the territory's 65,000 people are Polynesian people whose ancestors have occupied Samoa for at least 2,000 years. Their main language is Samoan, a Polynesian dialect, and many people speak English. Most American Samoans are Christians.

Most people live in villages, and their lives center around their families. Each family group is headed by a chief who represents the family in the village council, controls its property, and takes care of the sick or aged.

Economy

In 1961, when the United States began an economic development program in American Samoa, many people left their villages to take jobs in industries around Pago Pago. As part of the economic development program, thatch-roofed houses

Samoan children wear traditional as well as Western clothing. They attend new schools built with U.S. funds, and many are taught by television. Children from ages 6 to 18 must attend school.

were replaced by hurricane-proof concrete buildings, new schools were built, and teaching by television was introduced.

American Samoa's leading industry is tuna canning. Fish products make up more than 90 percent of all exports. The U.S. government has provided large amounts of money to help American Samoa develop a prosperous economy. Tourism has become an important source of income.

History and government

European explorers first reached Samoa in 1722, but there was little outside interest in the islands until the first mission was established in 1830. During the mid-1800's, two royal families ruled different parts of Samoa and fought over who would be king. Germany, the United Kingdom, and the United States supported rival groups.

In 1872, the Samoans agreed to let the United States use Pago Pago Bay as a naval coaling station. The United States was later given trading rights in the islands. In 1899, Germany, the United Kingdom, and the United

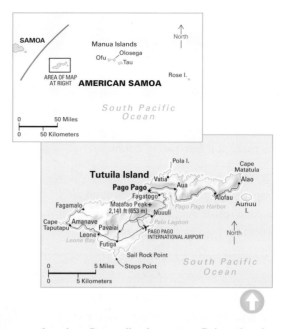

American Samoa lies in western Polynesia, about 2,300 miles (3,700 kilometers) southwest of Hawaii. The territory consists of seven islands, which have a total land area of 76 square miles (197 square kilometers).

Pago Pago, the capital and only urban center on American Samoa, has one of the best—and most beautiful—harbors in the South Pacific.

American-style football is a popular part of Samoan culture. Many Samoans have traveled to the United States and starred in both college and professional competition. These young players stand with their hands over their hearts as they listen to an inspirational speech by the territory's governor.

States signed a treaty dividing the islands between the United States and Germany. The U.S. Navy administered the U.S. islands until 1951, when they were transferred to the Department of the Interior. Afterward, the secretary of the interior appointed the governor of American Samoa.

In the early 1970's, the United States proposed that the territory elect its own governor, but the Samoans voted against the proposal three times. Many believed that the change would weaken their ties to the United States. American Samoans finally approved the proposal in 1976 and elected a governor in 1977.

Governors serve four-year terms. The territory's legislature has two houses. The Senate has 18 members chosen by county councils. The House of Representatives has 20 members elected by the people and one nonvoting delegate from Swains Island.

The U.S. Congress classifies American Samoa as an *unorganized and unincorporated territory*. Unincorporated territories are not eligible to become states and have fewer rights than incorporated territories. American Samoans elect a delegate to the United States House of Representatives. The delegate may vote in House committees, but not in House votes.

ANDORRA

Andorra is one of the smallest countries in the world, situated high in the Pyrenees Mountains between France and Spain. Andorra's unusual political history and interesting geography make it a fascinating country.

In the past, the steep, rocky mountains that surround Andorra made transportation and communication almost impossible. As a result, Andorra was cut off from the rest of the world for hundreds of years.

Visitors come to Andorra to enjoy the country's picturesque landscape, with old churches nestled in steep, rocky mountainsides. Tourism is the country's largest source of income, and Andorra's capital, Andorra la Vella, has become a leading tourist center. The banking industry is also important.

Early history

A Spanish ruler, the Count of Urgel, was the first known ruler of Andorra. He controlled the region in the A.D. 800's, until he gave it to the diocese of Urgel. In the 1000's, the bishop of Urgel asked a Spanish noble, the Lord of Caboet, to defend the region.

Later, a French noble, the Count of Foix, inherited the lord's duties through marriages. Soon, the French count and the bishop of Urgel were fighting over control of Andorra. They ended their differences by signing treaties in 1278 and 1288 that made them joint rulers.

Andorra is a parliamentary co-principality. Until 1993, the president of France and the bishop of Urgel, Spain, acted as co-princes of Andorra under the terms of the treaties signed in the 1200's. In 1993, the citizens of Andorra adopted their first constitution. The constitution made elected officials responsible for governing Andorra, and the role of the princes became largely ceremonial.

Way of life

Andorra's mountainous landscape limits agriculture. The main crop is tobacco, and cigarettes and cigars are among the country's main products. Other crops include oats, potatoes, and rye. Most of the mountain slopes are used for grazing sheep.

FACTS

Official name:	Principat d'Andorra (Principality of Andorra)
Capital:	Andorra la Vella
Terrain:	Rugged mountains dissected by narrow valleys
Area:	181 mi^2 (468 km^2)
Climate:	Temperate; snowy, cold winters and warm, dry summers
Main rivers:	Valira del Norte, Valira del Orient, Madriu
Highest elevation:	Coma Pedrosa, 9,665 ft (2,946 m)
Lowest elevation:	Riu Runer, 2,756 ft (840 m) below sea level
Form of government:	Parliamentary co-principality. Andorra has a democratic government led by the bishop of Urgel, Spain, and the president of France.
Head of state:	Two co-princes, the president of France and bishop of Seo de Urgel, Spain, who are represented locally by their representatives
Head of government:	Executive council president
Administrative areas:	7 parrouies (parishes)
Legislature:	Consell General de las Valls (General Council of the Valleys) with 28 members serving four-year terms
Court system:	Tribunal de Batlles (Tribunal of Judges); Tribunal de Corts (Tribunal of the Courts); Tribunal Superior de Justicia d'Andorra (Supreme Court of Justice of Andorra); lower courts
Armed forces:	France and Spain are responsible for Andorra's defense
National holiday:	Our Lady of Meritxell Day - September 8 (1278)
Estimated 2010 population:	84,000
Population density:	464 persons per mi^2 (179 per km^2)
Population distribution:	89% urban, 11% rural
Life expectancy in years:	Male, 80; female, 85
Doctors per 1,000 people:	3.6
Birth rate per 1,000:	10
Death rate per 1,000:	5
Infant mortality:	4 deaths per 1,000 live births
Age structure:	0-14: 15%; 15-64: 73%; 65 and over: 12%
Internet users per 100 people:	73
Internet code:	.ad
Languages spoken:	Catalan (official), French, Castilian, Portuguese
Religion:	Roman Catholic (predominant)
Currency:	Euro
Gross domestic product (GDP) in 2008:	N/A
Real annual growth rate (2008):	N/A
GDP per capita (2008):	N/A
Goods exported:	Electrical machines, food, vehicles
Goods imported:	Consumer goods, electrical machines
Trading partners:	France, Spain

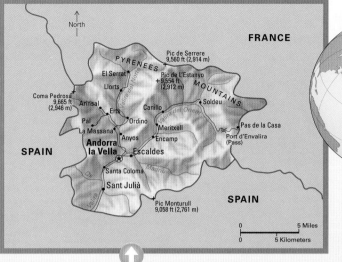

The tiny principality of Andorra is only about half the size of New York City. It lies on the south slope of the eastern Pyrenees between France and Spain.

The opening of roads to France and Spain in the 1930's, as well as the growth of tourism in the 1950's, changed some of the old ways that had been a part of Andorran life for hundreds of years. Many farmers and shepherds became shopkeepers and hotel owners, for example.

Despite the changes, life for most Andorrans still centers around the family, and political issues are still decided by family clans. Many Andorrans still live in big farmhouses with stone walls and rough slate roofs. Almost all of them are Roman Catholics, and religion greatly influences everyday life.

The wildlife of the Pyrenees enjoy one of the last areas of Europe that is free from human activity. The mountains form an effective barrier between people and nature, allowing many species to thrive.

1. Pyrenean ibex (goat)
2. Brown bear
3. Griffon vulture
4. Imperial eagle
5. Ptarmigan
6. Snow finch
7. Pyrenean desman
8. Alpine salamander
9. Turk's-cap lily
10. Pyrenean squill
11. Crocus
12. Pyrenean snakeshead
13. Scotch pine
14. Mountain pine

ANGOLA

The southwest African nation of Angola became independent in November 1975. Until that time, parts had been ruled by Portugal almost continuously since the early 1500's. From the 1960's to the early 2000's, the Angolans were plagued by violent revolution and civil war. A cease-fire was signed in 2002.

Early history

Angola's history started long before the Portuguese arrived on its shores. As early as 50,000 B.C., people are known to have lived in what is now Angola. Bantu-speaking groups settled there about 2,000 years ago. The Portuguese came in the early 1500's, and by the early 1600's they were taking great numbers of the local people as slaves for their colony in Brazil.

During the 1800's, the slave trade declined, and the Portuguese began to plant corn, sugar cane, and tobacco in Angola. In the late 1920's, after the Portuguese dictator António de Oliveira Salazar came to power, Portugal started to improve the region's economy, and thousands of Portuguese moved to Angola to start businesses.

Independence and civil war

During the 1950's, many Angolans began to demand freedom from Portuguese rule. In 1956, they organized the Popular Movement for the Liberation of Angola (MPLA). In 1961, MPLA members began a revolt that soon turned into a bloody guerrilla war. A Portuguese army, together with many Angolans, put down the uprising, but the rebels set up bases in neighboring countries. Other rebel groups also sprang up. In 1962, northern rebels formed the Front for the Liberation of Angola (FNLA). In 1966, southern rebels formed the National Union for the Total Independence of Angola (UNITA).

Portuguese military officers who had overthrown the Portuguese government gave Angola its independence in 1975. At first, the Angolans agreed to set up a government with representatives from all three rebel groups. However, each group wanted to head the government, and a civil war began. The FNLA and UNITA united against the MPLA, who received aid from the Communist countries of Cuba and the Soviet Union. When the MPLA finally won the war in April 1976, it formed a *Marxist*

FACTS

● Official name:	Republica de Angola (Republic of Angola)
● Capital:	Luanda
● Terrain:	Narrow coastal plain rises abruptly to vast interior plateau
● Area:	481,354 mi² (1,246,700 km²)
● Climate:	Semiarid in south and along coast to Luanda; north has cool, dry season (May to October) and hot, rainy season (November to April)
● Main rivers:	Cuanza, Cunene, Cuango
● Highest elevation:	Morro de Moco, 8,596 ft (2,620 m)
● Lowest elevation:	Atlantic Ocean, sea level
● Form of government:	Republic
● Head of state:	President
● Head of government:	President
● Administrative areas:	18 provincias (provinces)
● Legislature:	Assembleia Nacional (National Assembly) with 220 members serving four-year terms
● Court system:	Tribunal da Relacao (Supreme Court)
● Armed forces:	107,000 troops
● National holiday:	Independence Day - November 11 (1975)
● Estimated 2010 population:	18,484,000
● Population density:	38 persons per mi² (15 per km²)
● Population distribution:	57% urban, 43% rural
● Life expectancy in years:	Male, 40; female, 42
● Doctors per 1,000 people:	Less than 0.05
● Birth rate per 1,000:	47
● Death rate per 1,000:	21
● Infant mortality:	132 deaths per 1,000 live births
● Age structure:	0-14: 46%; 15-64: 52%; 65 and over: 2%
● Internet users per 100 people:	3
● Internet code:	.ao
● Languages spoken:	Portuguese, Bantu, other African languages
● Religions:	Roman Catholic 68%, Protestant 20%, Indigenous beliefs 12%
● Currency:	Kwanza
● Gross domestic product (GDP) in 2008:	$83.38 billion U.S.
● Real annual growth rate (2008):	13.2%
● GDP per capita (2008):	$4,816 U.S.
● Goods exported:	Mostly: crude oil Also: diamonds, fish, refined petroleum products
● Goods imported:	Food, machinery and electrical equipment, vehicles and vehicle parts
● Trading partners:	China, Portugal, South Africa, United States

Angola lies on the southwest coast of Africa. Cabinda, in the northwest, is part of Angola, even though it is separated from the rest of the country by the Congo River and a narrow strip of the Democratic Republic of the Congo.

government (one based on the ideas of German thinker Karl Marx, who developed the philosophy of Communism).

Angola's new government faced major problems. UNITA, aided by South Africa, continued to wage guerrilla war against it. Cuban troops helped the government fight the UNITA rebels. In 1988, South Africa stopped aiding the UNITA rebels. The next year, the government and UNITA announced a cease-fire. All Cuban troops were withdrawn by mid-1991.

In 1991, the MPLA legalized all political parties. In multiparty elections held in

1992, the leader of the MPLA, José Eduardo dos Santos, became president. However, UNITA protested that the elections were fraudulent, and civil war erupted once again. In 1994, the warring parties signed a peace treaty. However, in 1998, the agreement broke down. Another cease-fire ended the war in 2002.

In 2008, multiparty elections were held for the first time since 1992. The MPLA won the most seats in the National Assembly and dos Santos continued as president. In 2010, the Angolan government enacted a new constitution. The new constitution eliminated direct presidential elections. Instead, the head of the party with a majority of seats in the National Assembly becomes president.

Luanda is Angola's chief industrial center and port, as well as the country's capital and largest city. Ships transport giant containers and other cargo out of the harbor that opens into the Atlantic Ocean.

LAND AND PEOPLE

Angola forms part of the large inland plateau of southern Africa. The land gradually rises from the interior to the west, where it drops sharply to a narrow plain on the Atlantic coast.

Little vegetation grows on the coastal plain, but hilly grasslands cover most of the rest of the country. A rocky desert spans the south, while tropical forests cover the north.

Many rivers cross Angola. A few serve as waterways into the interior. Some, like the Cuango, flow north into the Congo River system. Others, like the Cuanza, flow west directly into the Atlantic Ocean.

Angola has 928 miles (1,493 kilometers) of coastline along the Atlantic. Railroads connect Angolan towns on the coast with the interior, providing an important link to the sea for neighboring Zambia and the Democratic Republic of the Congo. Luanda, the capital and largest city of Angola, is a major African seaport. Fishing is an important economic activity along the coast.

The economy of Angola is based largely on agriculture. Most of Angola's workers are employed in agriculture and farm or herd for a living. They often raise just enough food for their families. The main food crops are bananas, cassava, corn, and sugar cane. Some farmers grow coffee.

Mining is becoming increasingly important. Angola's land is rich in several mineral resources, especially diamonds, iron ore, and petroleum. Angola is one of the world's leading diamond producers. Most of the petroleum comes from Cabinda, the small Angolan district that lies to the northwest, separated from the rest of the nation by the Congo River and part of Congo. Petroleum accounts for almost all of Angola's exports.

Manufacturing is increasing in importance too. Angolan factory workers produce cement, chemicals, petroleum products, processed foods, and textiles.

At a diamond mine, tons of rock must be mined and crushed to uncover just one small stone. Diamonds are among the valuable mineral resources found in Angola. Angola also has vast deposits of iron ore and petroleum.

Dockside cranes tower over the Atlantic coast at Lobito, one of Angola's seaports. The city lies at the end of a railroad that has carried goods to and from neighboring Democratic Republic of the Congo as well as the interior of Angola. But Angola's civil war severely disrupted the railroad's operations.

Before Angola became independent, more than 400,000 Europeans and *mestizos* (people with both African and European ancestors) lived in Angola. Most fled the country after 1975, during the civil war between the government and rebel troops. Most of those who stayed live in Angola's urban areas and run small businesses or hold other jobs that require technical and management skills.

Angola's people belong to several different ethnic groups. The largest are the Ovimbundu, the Mbundu, the Kongo, and the Luanda-Chokwe.

Long ago, the Kongo people had a great kingdom that included part of Angola. Their capital, Mbanza, lay near what is now the northern Angolan town of Damba. The Kongo kingdom was weakened and eventually destroyed by the Portuguese slave trade.

Most of Angola's people, including the Kongo, speak languages that belong to the Bantu language group. Angolans of European descent, mestizos, and some others speak Portuguese, the official language. About 88 percent of the people are Christians, mostly Roman Catholics. Some other Angolans practice religions based on the worship of ancestors and spirits.

A young Angolan woman displays her finery—an elaborate beaded cap and a wealth of necklaces. Today, almost all the people of Angola are black Africans. Most Europeans fled the country during its civil war.

Fishery workers prepare their catch for market by drying it in the hot Angolan sun. Fishing is an important industry along Angola's coastline.

A rural settlement sprawls over an Angolan hillside among the grasslands and scattered trees of the inland plateau. About 40 percent of Angola's people live in rural areas and work as farmers and herders.

ANTARCTICA

Over 2,000 years before Antarctica was discovered, ancient Greek philosophers believed that a continent covered the southern end of Earth. Finally, this nearly barren land was sighted in 1820. During the mid-1800's, navigators sailed along its icy coast and learned that it was large enough to be called a continent. In the early 1900's, inland exploration began, and in 1911 the Norwegian explorer Roald Amundsen reached the continent's best-known location, the South Pole.

Antarctica is larger in area than either Europe or Australia. Nearly all of Antarctica is covered with thick ice sheets, one on either side of the Transantarctic Mountains. The icy layer, which averages approximately 7,100 feet (2,200 meters) thick, makes Antarctica the highest continent in terms of average elevation. It also increases Antarctica's surface area by filling deep basinlike areas of land that lie below sea level in West Antarctica. If the ice sheet melted, West Antarctica would become a group of islands.

The stormy waters of the Atlantic, Indian, and Pacific oceans isolate Antarctica from the other continents. Ships must steer around massive icebergs and break through huge ice sheets to reach the continent. On land, gigantic glaciers move slowly downhill toward the sea. Antarctica is the coldest, iciest region on Earth.

Temperatures in Antarctica rarely rise above 32° F (0° C). The world's lowest temperature, − 128.6° F (− 89.2° C), was recorded at Vostok Station in Antarctica on July 21, 1983. Strong, bitter winds also chill the air. Only a few small plants and insects can survive in Antarctica's dry interior. Its coastal waters, however, are rich in wildlife, and penguins, seals, and flying birds live or nest on the peninsula and offshore islands.

Antarctica is the site of more than one type of south pole. The south geographic pole, commonly referred to as the South Pole, is the point where all lines of longitude meet. It lies near the center of the continent. The south magnetic pole is the point indicated by compass needles. This location moves by as much as 5 to 10 miles (8 to 10 kilometers) in a year. In the early 2000's, it was off the coast of Wilkes Land.

Today, scientists from many countries maintain year-round research stations in Antarctica. Activities on the continent encourage international cooperation and the sharing of scientific knowledge. Several countries have claimed parts of the continent in the hope of controlling mineral resources found there, but the Antarctic Treaty places a freeze on existing claims and prohibits new ones.

ENVIRONMENT

The Antarctic ice sheets are two thick layers of ice and snow that together cover most of the continent. They formed from layers of snow that were pressed together over millions of years. High mountain peaks and a few bare rocky areas are the only visible land. Underneath the ice, however, Antarctica has mountains, lowlands, and valleys—much like the landforms of other continents.

The Antarctic ice sheets form the largest body of fresh water or ice in the world. With a volume of 7-1/4 million cubic miles (30 million cubic kilometers), it represents about 70 percent of the world's fresh water. If the ice sheets melted, Earth's oceans would rise and flood coastal cities around the world.

The Transantarctic Mountains cross the entire continent. Several ranges make up the Transantarctic chain. Some peaks rise more than 14,000 feet (4,300 meters). Two large gulfs cut into the continent at opposite ends of the mountains, smaller bays indent the coastline, and channels separate offshore islands from the mainland.

Broad, flat, floating parts of ice sheets called

Antarctica lies beneath a layer of ice and snow that measures up to 11,500 feet (3,500 meters) in the thickest areas. The Transantarctic Mountains divide the continent into two regions: East Antarctica and West Antarctica. The South Pole lies near the center on an icy, windswept plateau. The southern parts of the Atlantic, Indian, and Pacific oceans meet at Antarctica to form a body of water often called the Antarctic Ocean or Southern Ocean.

• Year-round research station

Mount Erebus is Antartica's most active volcano. It is on Ross Island, which is on the side of West Antarctica that faces New Zealand. The volcano rises 12,448 feet (3,794 meters). Erebus has been erupting almost continously at least since the early 1970's.

Ice-free areas called *dry valleys* appear where Antarctica's glaciers have retreated and wind prevents snow from collecting. About 98 percent of the continent lies beneath ice and snow.

ice shelves fill several of Antarctica's bays and channels. The largest one is the Ross Ice Shelf, which measures about 2,300 feet (700 meters) thick at its inner edge. In summer, the outer edges of these ice shelves break away and form immense, flat icebergs. Some of these icebergs have measured as much as 5,000 square miles (13,000 square kilometers) in area.

The Transantarctic Mountains divide Antarctica into two natural land regions—East Antarctica and West Antarctica.

East Antarctica, more than half the continent, consists of rocks that are more than 570 million years old. Mountains, valleys, and glaciers line the coast. The interior part of East Antarctica is a plateau about 10,000 feet (3,000 meters) above sea level. The South Pole lies on the plateau, at the center of the continent. This pole is Earth's southernmost point, where all lines of longitude meet. East Antarctica also has the *south magnetic pole,* the southern point indicated by compass needles.

West Antarctica, which borders the Pacific Ocean, contains little of the old rock of East Antarctica. West Antarctica developed later as part of the Ring of Fire, a string of volcanoes that encircles the Pacific Ocean. The region includes several mountain ranges and volcanoes. Vinson Massif, the highest point in Antarctica at 16,067 feet (4,897 meters), stands in the Ellsworth Mountains. Mount Erebus, Antarctica's most active volcano, towers 12,448 feet (3,794 meters) above Ross Island.

The Antarctic Peninsula is a mountainous, *S*-shaped finger of land that points out from West Antarctica toward South America. The peninsula is actually a continuation of the Andes Mountain chain of South America.

WILDLIFE

Many millions of years ago, Antarctica was an ice-free continent. Scientists have found fossils of trees, dinosaurs, and small mammals that once lived there. Today, only a few small plants and insects can survive in Antarctica's dry interior. Most land animals live at the edges of the continent. The continent's largest land animal is a wingless *midge,* a type of fly no more than 1/2 inch (12 millimeters) long. To avoid freezing to death, some lice, mites, and ticks cling to mosses, the fur of seals, or the feathers of birds.

Few plants grow in Antarctica's forbidding, ice-covered land and harsh climate. Mosses, the most common Antarctic plants, cling to rocky areas, mostly on the coasts. Only two flowering plants grow in Antarctica, both on the northern part of the Antarctic Peninsula. One is a grass that forms dense mats on sunny slopes, while the other, an herb, grows in short, cushionlike bunches.

Fossils of ferns from the Triassic Era show that Antarctica was once an ice-free continent with more types of green plants than the mosses common today. Mosses cling to rocky areas, mostly on the coasts

Curious crabeater seals cluster around a diver under a hole in the ice. While many kinds of seals are slow and clumsy on land or ice, the crabeater seal can move at a speed of about 15 miles (24 kilometers) per hour—almost as fast as a person can run.

A Weddell seal nuzzles her pup. Seals have few enemies besides people and killer whales. To escape whales, Weddell seals of the Antarctic can dive as deep as 2,360 feet (719 meters) and stay underwater as long as 43 minutes.

Life in the ocean

The Antarctic Ocean has abundant wildlife. The most common ocean animal is the *krill,* a small, shrimplike creature that feeds on tiny floating organisms. Many other Antarctic animals depend on krill for food. Many Antarctic animals also eat *squid*—a soft, boneless sea animal. In addition, about 100 kinds of fish live in the ocean, including Antarctic cod, icefish, and plunderfish.

Several kinds of whales migrate to Antarctica for the summer. The blue whale—the largest animal that has ever lived—is one of these. This rare mammal feeds on krill and grows up to 100 feet (30 meters) long. Humpback whales and killer whales are among the other kinds of whales that spend summers in Antarctica.

Various kinds of seals also live in Antarctica, spending most of their lives in the water, where they swim, dive, and catch food. Most of them nest on the coasts, but the Antarctic fur seal nests on nearby islands. The southern elephant seal, the largest seal in the world, also lives in the Antarctic region. These seals have large noses and tough skin, and the males may reach a length of 16 feet (5 meters).

Birdlife

Several kinds of penguins breed on the continent. On land, these flightless birds waddle awkwardly, but in water they are swift, skillful swimmers that feed on fish and other food they find in the ocean.

Adélie penguins, the most common kind, build nests of pebbles on the coasts. Other inhabitants of the mainland are the quieter emperor penguins. After the female emperor lays an egg on the ice, the male rests the egg on his feet to warm it. Chinstrap, gentoo, king, and macaroni penguins nest on the Antarctic Peninsula and on islands. Rockhopper penguins nest only on islands north of Antarctica.

More than 40 kinds of flying birds spend the summer in Antarctica. Birds that nest on land but spend most of their time diving for food include albatrosses, prions, and a large group of sea birds known as petrels. Other birds, such as cormorants and gulls, return to land more frequently. Some steal food from the nests of other birds.

Emperor penguins are the world's largest penguins, growing to about 4 feet (1.2 meters) tall. Young emperors huddle under the body of an adult bird to keep warm.

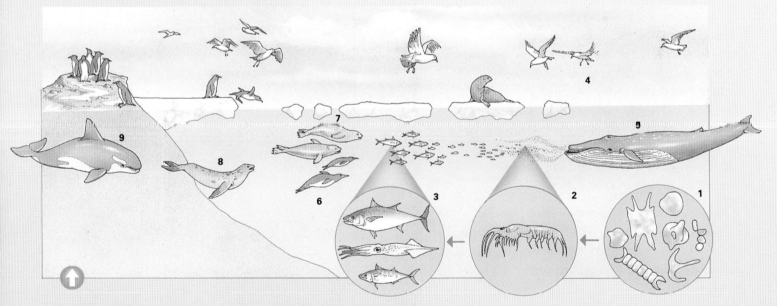

The Antarctic food chain nourishes a great variety of wildlife. Tiny floating organisms known as plankton (1) support krill (2), a small, shrimplike creature that is the most common animal of the Antarctic Ocean. Krill is rich in protein and a key source of food for many varieties of fish (3), birds (4), and larger creatures such as fin whales (5). Fish, in turn, provide food for penguins (6) and several species of seal (7). Leopard seals (8) hunt other seals as well as penguins. Killer whales (9) also hunt seals, penguins, and smaller whales.

PROBLEMS AND CHALLENGES

Scientists believe that Antarctica originally belonged to a land mass that included Africa, Australia, India, and South America. By about 140 million years ago, the land had begun to break apart. The parts gradually drifted to their present locations, and Antarctica became a separate continent. Evidence for this theory of *continental drift* comes from studying such things as mountain systems, fossils, the location of earthquakes and volcanoes, and magnetism in ancient rocks.

Antarctica has a variety of mineral resources. However, most of Antarctica's mineral deposits are too small to be mined efficiently. Icebergs, rough waves, and strong winds hamper drilling operations at sea. In addition, many scientists fear that large-scale mining would harm Antarctica's environment.

During the International Geophysical Year (July 1, 1957, to Dec. 31, 1958), 12 countries established more than 50 scientific stations on Antarctica and nearby islands. Seven of those countries have claimed parts of Antarctica, but the other five countries do not recognize Antarctic claims.

In 1959, officials of the 12 countries signed the Antarctic Treaty. The treaty, which took effect in 1961, allows people to use Antarctica for peaceful purposes only, such as exploration and scientific research, and requires scientists to share any knowledge that results from their studies. The treaty forbids military forces to enter Antarctica, except to assist scientific expeditions, and outlaws the use of nuclear weapons and the disposal of radioactive wastes in Antarctica.

Since the Antarctic Treaty took effect, several other countries have signed the document and set up scientific programs in Antarctica. Members have also added laws that protect Antarctic plants and animals.

In 1991, the Antarctic Treaty nations signed the Madrid Protocol. This agreement, which went into effect in 1998, defines Antarctica as a natural reserve devoted to peace and science. The Madrid Protocol prohibits mineral exploitation in Antarctica and establishes strict rules designed to protect the Antarctic environment.

Today, more than 40 year-round scientific stations operate on the continent and nearby islands. Some Antarctic studies concern all the continents. Significant research deals with *ozone,* a form of oxygen. A layer of ozone surrounding Earth protects all living things from certain harmful rays of the sun. In the mid-1980's, scientists discovered a "hole" in the ozone layer above Antarctica.

A scientist takes magnetic readings near an Antarctic base. Researchers there have studied such topics as earthquakes, gravity, magnetism, oceans, and solar activity. Others have measured the icecap's thickness.

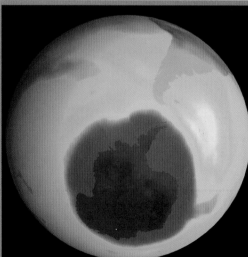

The "hole" in Earth's ozone layer over Antarctica shows up in a satellite image. The ozone layer protects living things from certain harmful rays of the sun.

Researchers also are studying the melting of the ice sheets. This research helps them predict how much sea levels might rise if Earth's average temperature continues to increase.

Since the late 1900's, access to Antarctica has become easier, and tourism has grown. Researchers face challenges in managing the Antarctic environment as the human population increases.

In 2007, scientists from dozens of nations launched a study of the Arctic and Antarctic called the International Polar Year 2007-2008. The investigation involved more than 200 research projects, with a focus on how climate change affects the polar regions and their inhabitants.

McMurdo Station, the U.S. research base on Ross Island has Antarctica's largest community. About 1,000 scientists, pilots, and other specialists live there each summer. Only about 250 people stay at the station for the winter, which lasts from May through August.

In February, 2010, during the Antarctic summer, a large portion of the Mertz Glacier, called the Mertz Glacier Tongue, broke away from the main glacier in a process called *calving*. The result of the calving was the formation of an iceberg that reportedly was 48 miles (78 kilometers) long and 24 miles (39 kilometers) wide. It had a mass of between 700 and 800 billion tons.

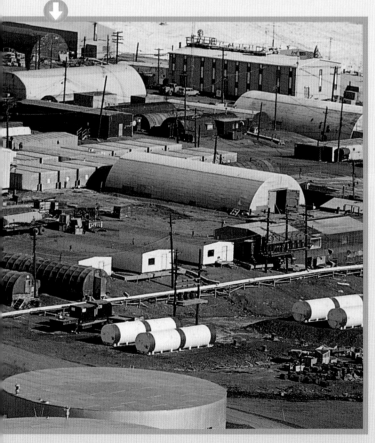

ANTIGUA AND BARBUDA

Antigua and Barbuda is an island country in the Caribbean Sea, at the northern end of the group of islands called the Lesser Antilles. The nation consists of three islands: Antigua, Barbuda, and Redonda. Most of the islanders are descendants of Africans, while people of European and mixed ancestry form a small minority.

Antigua and Barbuda has a total population of about 88,000. About 98 percent of the people live on Antigua and 2 percent on Barbuda.

When Christopher Columbus discovered Antigua in 1493, the island was inhabited by Carib Indians. The Spaniards killed many of the Caribs and forced others to work in the mines of Hispaniola. British settlers established a colony on Antigua in 1632, and later they also colonized Barbuda and Redonda. All three islands became known as the colony of Antigua.

The British settlers brought Africans to the islands to work as slaves on sugar cane plantations. The slaves were freed in 1834, the year after the United Kingdom abolished slavery throughout its empire. Most of the British people eventually left, but the United Kingdom retained control of the colony.

In 1967, Antigua became part of the West Indies Associated States and gained control of its internal affairs. On Nov. 1, 1981, it became the independent nation of Antigua and Barbuda.

Tourism and the economy

Tourism is the main economic activity on the islands, and the tourist industry employs many of the people. The country's beautiful beaches and sunny climate attract tourists from the United Kingdom, the United States, and many other countries.

Agriculture is also important to the country's economy. Farmers raise cotton, fruit, livestock, sugar cane, and vegetables. Droughts often damage the country's

FACTS

Official name:	Antigua and Barbuda
Capital:	St. John's
Terrain:	Mostly low-lying limestone and coral islands, with some higher volcanic areas
Area:	171 mi² (442 km²)
Climate:	Tropical marine; little seasonal temperature variation
Main rivers:	N/A
Highest elevation:	Boggy Peak, 1,319 ft (402 m)
Lowest elevation:	Caribbean Sea, sea level
Form of government:	Constitutional monarchy
Head of state:	British monarch, represented by governor general
Head of government:	Prime minister
Administrative areas:	6 parishes and 2 dependencies
Legislature:	Parliament consisting of the Senate with 17 members and the House of Representatives with 17 members serving five-year terms
Court system:	Eastern Caribbean Supreme Court
Armed forces:	170 troops
National holiday:	Independence Day (National Day) - November 1 (1981)
Estimated 2010 population:	88,000
Population density:	515 persons per mi² (199 per km²)
Population distribution:	70% urban, 30% rural
Life expectancy in years:	Male, 72; female, 76
Doctors per 1,000 people:	0.2
Birth rate per 1,000:	17
Death rate per 1,000:	6
Infant mortality:	18 deaths per 1,000 live births
Age structure:	0-14: 28%; 15-64: 65%; 65 and over: 7%
Internet users per 100 people:	74
Internet code:	.ag
Languages spoken:	English (official), local dialects
Religions:	Protestant 81.8%, Roman Catholic 10.4%, other 7.8%
Currency:	East Caribbean dollar
Gross domestic product (GDP) in 2008:	$1.20 billion U.S.
Real annual growth rate (2008):	2.1%
GDP per capita (2008):	$14,313 U.S.
Goods exported:	Machinery, petroleum products, transportation equipment
Goods imported:	Food and live animals, machinery, petroleum products, transportation equipment
Trading partners:	Caribbean Islands, United Kingdom, United States

English Harbour at St. John's, Antigua was formerly a British naval dockyard. Today, it is a popular port of call for cruise ships and yachts. Tourism is the backbone of the economy of Antigua and Barbuda, but the government also encourages the development of small industries.

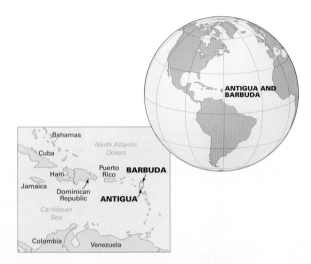

Antigua is the largest of the three islands that make up the Caribbean nation of Antigua and Barbuda. This island country lies about 430 miles (692 kilometers) north of Venezuela and covers 108 square miles (280 square kilometers). About 98 percent of the people live on Antigua, and the remaining 2 percent live on Barbuda. Redonda, the third island in the group, is an uninhabited rocky islet.

An Antiguan cook prepares lobsters as tourists watch. The teeming underwater life of lagoons and coral reefs makes Antigua and Barbuda a paradise for divers and snorkelers.

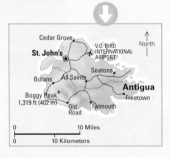

The island of Barbuda lies north of Antigua and covers 62 square miles (161 square kilometers). Much of the land is a nature preserve, where birds, turtles, and lizards thrive.

crop production. As a result, the government encourages the development of small industries to strengthen the economy. The country's manufactured products include clothing and household appliances.

Environment

The islands of Antigua and Barbuda are volcanic in origin, but erosion by wind and rain has worn down the volcanoes. The islands are now mostly flat. Beautiful white sandy beaches line the coasts.

Antigua also has a number of bays and inlets. St. John's, situated on the northwest coast of the island, is the capital and largest city. It is a famous port of call for cruise ships.

Barbuda, less developed than Antigua, is mainly a nature preserve. Unlike many islands in the West Indies, Barbuda still has abundant animal life. The island provides a haven for numerous species of birds, lizards, and turtles. Barbuda also has a large colony of rare frigate birds, and turtles lay their eggs on the beaches. Many kinds of fish swim in the lagoons, which lie within protective coral reefs.

ARGENTINA

Argentina is a huge, sprawling land that occupies most of the southern part of the South American continent. It is the second largest country in South America in area and in population; only Brazil is larger.

Argentina's landscape ranges from lofty mountains and arid deserts to vast plains and dense forests. Argentina is a country of many contrasts—not only in its geography, but also in its people and their way of life.

From the rolling, grass-covered plains of the north to the wild and windswept plateaus of the south, Argentina has a rugged, natural beauty all its own. The craggy, snow-covered Andes Mountains rise up along its western border with Chile, while the South Atlantic Ocean meets its eastern shoreline.

Parts of Argentina are parched with drought most of the year. But along the Brazilian border, the thundering waters of Iguaçu Falls plunge 237 feet (72 meters) down a series of ledges in one of the world's most spectacular sights.

Most Argentines are of Spanish or Italian ancestry. The *indigenous* (native) population—descended from the groups that lived in the region before the arrival of Spanish settlers in the 1500's—is small compared to that of other Latin American countries. Indigenous Argentines live mainly in isolated areas of northern and southwestern Argentina.

Like the land around them, the Argentines are hardy, rugged people. On the huge ranches of the Pampas, *gauchos* (cowboys) herd cattle in the proud tradition of their ancestors. In Buenos Aires, the elegant Colón Theater—one of the world's finest opera houses—draws large, enthusiastic audiences. Argentine families enjoy seaside vacations along the white, sandy beaches of the country's Atlantic coast.

ARGENTINA TODAY

Argentina has enjoyed greater social and economic development than many other South American countries. It has a large middle class and an advanced level of industrialization. But economic decline, political instability, social conflict, and violence hampered Argentina's progress during the last half of the 1900's.

Politically, the country has suffered through unstable governments, harsh military dictatorships, and political terrorism. Since 1930, military dictatorships have often ruled the country, and parts of the Constitution were suspended in 1976 when the military overthrew President Isabel Perón's government.

Isabel Perón was the third wife of Juan Perón, who had served as Argentina's president between 1946 and 1955 and again from October 1973 until his death in July 1974. Isabel Perón became president when he died.

Argentina's problems began to increase after Isabel Perón took office. The inflation rate soared. Political extremists engaged in terrorist attacks. In 1976, military leaders arrested Perón, dissolved the Congress, and took control of the government.

Military rule

The reign of terror that followed Perón's removal from office became known as the "dirty war." To the military, it was an attempt to rid the country of leftist opposition and political terrorism. For Argentina's people, however, it was a dark period of violence. The military not only restricted civil rights but also imprisoned, tortured, and killed thousands.

In 1982, a war with the United Kingdom over control of the Falkland Islands further damaged the Argentine economy and led to serious political unrest. In 1983, Raúl Alfonsín was elected president, and civilian government was reestablished. Alfonsín promised an investigation into the actions of previous governments. Three former presidents and several other officials were convicted and sentenced to prison for their involvement in murders and torture.

In 1989, Carlos Saúl Menem was elected president. He granted pardons to some of the people convicted in the mid-1980's for their involvement in murders and torture.

FACTS

Official name:	Republica Argentina (Argentine Republic)
Capital:	Buenos Aires
Terrain:	Rich plains of the Pampas in northern half, flat to rolling plateau of Patagonia in south, rugged Andes along western border
Area:	1,073,519 mi² (2,780,400 km²)
Climate:	Mostly temperate; arid in southeast; subantarctic in southwest
Main rivers:	Paraná, Uruguay, Negro, Salado, Colorado, Bermejo
Highest elevation:	Cerro Aconcagua, 22,835 ft (6,960 m)
Lowest elevation:	Valdés Peninsula, 131 ft (40 m) below sea level
Form of government:	Republic
Head of state:	President
Head of government:	President
Administrative areas:	23 provincias (provinces) and 1 distrito federal (federal district)
Legislature:	Congreso Nacional (National Congress) consisting of the Senate with 72 members serving six-year terms and the Chamber of Deputies with 257 members serving four-year terms
Court system:	Corte Suprema (Supreme Court)
Armed forces:	76,000 troops
National holiday:	Revolution Day - May 25 (1810)
Estimated 2010 population:	40,519,000
Population density:	38 persons per mi² (15 per km²)
Population distribution:	91% urban, 9% rural
Life expectancy in years:	Male, 72; female, 79
Doctors per 1,000 people:	3.0
Birth rate per 1,000:	18
Death rate per 1,000:	8
Infant mortality:	13 deaths per 1,000 live births
Age structure:	0-14: 26%; 15-64: 64%; 65 and over: 10%
Internet users per 100 people:	28
Internet code:	.ar
Languages spoken:	Spanish (official), Italian, English, German, French
Religions:	Roman Catholic 70%, Protestant 9%, Muslim 1.5%, Jewish 0.8%, other 18.7%
Currency:	Argentine Peso
Gross domestic product (GDP) in 2008:	$331.19 billion U.S.
Real annual growth rate (2008):	7.1%
GDP per capita (2008):	$8,333 U.S.
Goods exported:	Agricultural products, cooking oil, petroleum and natural gas, processed foods
Goods imported:	Chemicals, machinery and mechanical equipment, metal products, plastics, transportation equipment
Trading partners:	Brazil, Chile, China, Germany, Mexico, Spain, United States

Soldiers at Argentina's presidential residence in Buenos Aires wear a traditional ceremonial uniform in the style of the 1800's.

Problems continue

Menem was reelected in 1995. During his second term, the government borrowed heavily, and its debts increased. Domestic interest rates rose, many companies closed, and many workers lost their jobs. By the late 1990's, Argentina was in a recession.

Fernando de la Rua was elected president in 1999. In 2000, his government increased taxes and made massive spending cuts. In late 2001, many people feared that the government would reduce the value of the peso, so they rushed to banks to withdraw money and convert their pesos to dollars. In response, the government limited the amount people could withdraw each month from their accounts. Violent protests broke out. Soon, the president and his cabinet resigned.

An unsettled period followed, during which three other leaders served as president, holding office for only hours or days. In 2002, Argentina's Congress chose Eduardo Duhalde as president. The economy stabilized, and by early 2003, the government had ended limits on withdrawals.

In 2003, Néstor Kirchner became president. He worked to reorganize the police and armed forces, end corruption, review past human rights abuses, and negotiate terms for repayment of the foreign debt.

Argentina is the second largest country in South America in area and population. The name Argentina comes from argentum, the Latin word for silver. The Spanish conquistadors who arrived in the 1500's believed that Argentina had large deposits of silver and gold.

In 2007, Argentine first lady Cristina Fernández de Kirchner, who had been a senator for more than 10 years, was elected president. She promised to continue her husband Néstor Kirchner's left-leaning policies. Néstor Kirchner died in 2010. Cristina Kirchner was reelected to a second term in 2011.

ENVIRONMENT

Argentina is the eighth largest country in the world. It shares the southern part of the South American continent with Chile, its neighbor to the west. Bolivia, Paraguay, Brazil, and Uruguay also border Argentina. Because Argentina covers such a vast expanse of land, its climate and geography range from hot and humid plains in the northeast to a bare, wind-swept plateau in the south. The country's four main land regions are Northern Argentina, the Pampas, the Andine, and Patagonia.

Northern Argentina

Northern Argentina is a huge lowland plain that lies east of the Andes Mountains and north of the Córdoba Mountains. The Paraná River divides Northern Argentina into two parts—the Gran Chaco, or Chaco, and Mesopotamia.

Few people live in the Chaco. Much of the region is covered by forests, and harvesting the quebracho tree is the main economic activity. The quebracho yields *tannin,* a chemical used in the leather industry, and the very hard wood of the quebracho tree is used to make telephone poles and railroad ties.

The Chaco has drought conditions most of the year, but heavy rains fall during the summer, causing riverbeds to overflow. Farmers plant mainly corn, cotton, soybeans, sunflowers, and wheat.

Argentina's climate is as varied as its geography. The tropical north has high temperatures, with plentiful rainfall in the northeast, while most of Patagonia receives less than 10 inches (25 centimeters) of rain a year.

A spectacular wall of ice on the Perito Moreno glacier on Lake Argentino sparkles in the sunlight. Perito Moreno is located in the magnificent Los Glaciares National Park in the southern Argentine Andes.

Like its ancient Middle Eastern namesake, Mesopotamia lies between two rivers. Argentina's Mesopotamia—also known as *Entre Ríos* (Between Rivers)—lies between the Paraná and the Uruguay. Mesopotamia is a region of fertile, grass-covered plains where farmers graze cattle, sheep, and horses. They also grow citrus fruits, corn, soybeans, rice, and wheat. The holly tree grows wild in Mesopotamia, and its dried leaves are used to make the tea known as *maté*—Argentina's national beverage.

The Pampas

The Pampas is a fertile plain that fans out around Buenos Aires, extending south and west all the way to the Andean foothills. It makes up about one-fifth of Argentina's total area. This vast, seemingly endless plain boasts some of the world's richest soil. Fields of alfalfa, corn, and wheat cover much of the land. In the drier western Pampas, huge herds of cattle graze on large *estancias* (ranches).

About three-fifths of Argentina's people live in the Pampas. In addition to Buenos Aires, with a population of about 14 million in its metropolitan area, several other large cities stand in or near the Pampas.

The Andine and Patagonia

The Andine, the mountainous western region of Argentina, is made up of two subregions—the Andes and the Piedmont. The Andes, which separate Argentina from Chile, include Aconcagua, the highest mountain in the Western Hemisphere at 22,835 feet (6,960 meters). In the northern part of the Argentine Andes, an area of plateaus called the Puna provides grazing land. The south consists mainly of snow-covered peaks and sparkling lakes.

East of the Andes lies the Piedmont, a region of low mountains and desert valleys. Mountain streams provide water for irrigation, making the Piedmont suitable for growing such crops as alfalfa, corn, cotton, grapes, and sugar cane.

Patagonia is a dry, windswept plateau in southern Argentina. Few people live there. Sheep-raising is the chief occupation. In the northern part of the region, farmers raise some fruits in river valleys. The islands of Tierra del Fuego lie off the southern tip of South America.

Grassy plains extend as far as the eye can see on a large ranch in Northern Argentina's Mesopotamia region. Ranchers graze cattle, sheep, and horses on Mesopotamia's vast pastureland, one of the leading wool-producing regions in Argentina.

A small settlement on the banks of the Chubut River stands out against the barren landscape of Patagonia. Although Patagonia covers more than one-fourth of Argentina, only about 5 percent of the people live there.

PEOPLE

About 250,000 Indians may have lived on the land that is now Argentina when the first Europeans arrived in the 1500's. By 300 years later, however, the Indian population had greatly declined. During the Spanish occupation, many Indians were killed by the Europeans, and many others died from diseases brought by the settlers. Some intermarried with the Europeans and produced a mestizo population.

Today, Argentina has a small Indian population compared with other South American countries—only about 50,000 Indians of unmixed ancestry live there. Most of the Indians have settled in isolated areas, such as the Andes, the Gran Chaco, and Patagonia.

Most of Argentina's people are of European ancestry, chiefly Italian and Spanish. Mestizos make up most of the rest of the population. European immigration has had an important effect on the growth of Argentina's population.

During the mid-1800's, only about a million people lived in the area known as the United Provinces of the Río de la Plata—or simply the United Provinces of La Plata. This area included much of the region that is now Argentina, except Patagonia. The United Provinces of La Plata had declared its independence from Spain in 1816, and named itself Argentina in 1860.

To encourage settlement of the country's vast interior, the leaders of the new republic developed an immigration policy that attracted Europeans wishing to build a new life in a new land. As a result, over the next several decades, huge waves of immigrants increased the population dramatically. By 1914, Argentina's population had reached nearly 8 million.

Most of the immigrants between 1860 and 1930 came from Italy and Spain, but Argentina also attracted settlers from such countries as Argentina, France, Germany, Portugal, Russia,

Switzerland, and the United Kingdom. After 1930, many immigrants came from Eastern Europe—specially Poland—and from the Middle East.

Spanish is Argentina's official language. But because such a large percentage of the Argentine population is of European ancestry, many people speak a second European language in addition to Spanish. Also, many Argentines read one of the foreign-language newspapers published daily in Buenos Aires.

Most European immigrants settled in the cities, where greater job and educational opportunities enabled the newcomers to enter the middle class. Today, middle-class people in Argentina—a group that is larger than the middle class of most other Latin American countries—enjoy a comfortable life and adequate diet.

⬆
Buenos Aires' Colón Theater presents fine opera, ballet, and concerts. The theater's large, enthusiastic audiences include many of Argentina's affluent upper-class and middle-class people.

The Andes form a backdrop for one of the world's southern-most settlements—Ushuaia, on the island of Tierra del Fuego. Although only about 11,000 people live in Ushuaia, the growing popularity of trips to Antarctica has made the island a tourist center.

Argentines are noted for their love of meat, especially beef. Barbecues are a favorite event, and Argentines especially enjoy *asado con cuero,* in which beef is roasted in its hide over an open fire. *Pucheros* (stews of chicken or other meat with vegetables) and *empanadas* (pastries stuffed with meat or seafood, eggs, vegetables, and fruit) are also popular.

Argentines also enjoy many religious festivals throughout the year. About 70 percent of the people are Roman Catholics, and their colorful processions and fireworks celebrate important church holidays.

During the *carnival,* a festival held before Lent, Argentines dance in the streets. In addition to religious festivals, their merrymaking includes wine and beer festivals and corn and wheat festivals. Festivals of folklore and music are also frequent events in many parts of the country.

River Plate Stadium in Buenos Aires is home to the popular River Plate Athletic Club soccer team. Soccer is Argentina's most popular team sport, and soccer games draw thousands of fans. Built in 1938, the stadium holds more than 65,000 people and is also a popular concert venue.

A street vendor in the city of La Plata barbecues beef and sausages for his customers. The Argentine diet emphasizes meat, especially beef, and some Argentines eat beef at every meal.

GAUCHOS

Out on the Argentine frontier, where the prairies of the Pampas stretch out as far as the eye can see, large herds of cattle graze peacefully under the bright afternoon sun. Suddenly, the sound of hooves breaks the silence, and a lone figure on horseback appears in the distance.

Sitting high in the saddle, the rider wears the colorful woolen poncho that has become his trademark, and he carries a *bola*—a long cord with stone balls tied at the end. Holding the horse's reins with one hand, he twirls the bola over his head with the other. The twirling bola makes a whooshing sound as he rides deep into the herd. The gaucho throws the bola so that it winds around and entangles the animal at which it is aimed.

He is a freedom-loving, rough-and-tumble *gaucho*—one of the South American cowboys made famous through story, song, and literature. Today, the gauchos work mainly as extra hands on Argentina's *estancias* (cattle ranches), but in the 1800's, the gauchos roamed the Pampas, living a wild and independent life that has come to symbolize the essence of freedom.

For 200 years, the gaucho way of life has captured the Argentine imagination. The spirit of the gaucho lives on, not only in folk tales, but also in the works of Argentina's greatest writers. The poet José Hernández wrote about the gauchos in his epic *Martín Fierro:* "Dead, the gaucho still survives—in the literature he inspired . . . and in the blood of every Argentine."

The gaucho also influenced other Argentine arts. The country's first important painter, Prilidiano Pueyrredón, created popular gaucho scenes during the 1800's. During the early 1900's, the classical composer Alberto Ginastera drew upon gaucho songs and dances in his works.

The first gauchos

While Hernández's poem *Martín Fierro* portrays the gauchos as heroes, describing their lonely life on the plains and their battles with Indians and with a

Performing in a rodeo, a gaucho displays the horsemanship for which the gauchos have been famous since the 1800's. In the past, gauchos lived off the land and spent most of their money on silver spurs, silver belts, and dazzling ornaments for their horses.

A gaucho in Buenos Aires shows off the traditional costume of these fiercely independent cowboys. A brightly colored poncho is worn over a cloth jacket, with a silver belt and facon (knife) at the waist. A wide-brimmed, cowhide hat completes the outfit.

cause they were skilled horsemen and familiar with the countryside, they were very successful in avoiding the authorities.

The gauchos caught wild cattle and sold hides in illegal trade on the Brazilian frontier. At night, they gathered around roaring fires for *asados* (barbecues of fresh beef). While drinking *maté* (tea) from a gourd, they swapped stories of their adventures and sang haunting melodies to the strum of a guitar.

Modern gauchos

The gaucho way of life came to an end in the late 1800's with the development of refrigerator ships. The ability to export meat made cattle-raising a big business, and a growing number of cattle ranches began to appear on the frontier. Finally, what was once a vast wilderness became the fenced-off property of wealthy *estancieros* (farmers).

Today, the descendants of the original gauchos display their legendary rope-throwing skills and horsemanship at rodeos. With each performance, they celebrate the proud, independent tradition born long ago on the Argentine frontier, and the gaucho spirit remains alive.

⬆

A spirit of rugged independence is seen on a gaucho's face. The origins of the word gaucho are unknown, but some people trace it to the Arawakan Indian word cachu, meaning comrade.

government that did not understand them, history tells a somewhat different story. The forerunners of the gauchos—mestizos who roamed the plains during the 1600's—were thought of as troublemakers and horse thieves. Often desperately poor, they wandered the frontier on horseback, spending most of their time in the saddle and sleeping under the stars.

In those days, thousands of semiwild cattle that had escaped the Spanish settlements grazed on the Pampas. These cattle provided fresh meat for the gauchos, and water from the rivers was readily available. The abundance of food and drink allowed the gauchos to live freely on the wild frontier, following no law but their own. Be-

A modern-day gaucho uses a bola to catch a fleeing calf. When thrown, the bola coils around an animal's legs, bringing it to the ground.

BUENOS AIRES

Buenos Aires, the capital and by far the largest city in Argentina, is also the nation's chief port and industrial center. The city lies along the southern shore of a wide, funnel-shaped bay called the *Río de la Plata* (Silver River).

Buenos Aires is one of the largest metropolitan areas in the world, and about a third of Argentina's people live there. The city itself covers 78 square miles (203 square kilometers), and the metropolitan area spreads over 1,479 square miles (3,830 square kilometers).

Despite its enormous size and huge population, Buenos Aires has an air of spaciousness and tranquillity. Only a few skyscrapers dot the skyline, while numerous parks and plazas line the city's broad avenues. The world's widest street, the Avenida 9 de Julio, runs through the central business district. The Avenida 9 de Julio is 425 feet (130 meters) wide and divided into three smaller streets by grassy strips.

The people of Buenos Aires proudly call themselves *porteños* (port dwellers). Most of the porteños are of Spanish or Italian ancestry. Other groups are descended from English, French, German, Lebanese, Polish, Russian, Jewish, and Syrian immigrants. Many of these immigrants were part of the huge wave of European immigration during the late 1850's.

A cafe owner in Buenos Aires prepares to open for business. Porteños, as the citizens of Buenos Aires are called, enjoy chatting with friends in the many cafes that line the city's streets and boulevards.

Brightly painted houses are typical of the barrio (residential neighborhood) of La Boca. Every Sunday, local artists gather to display their work along Caminito Lane in La Boca.

The Casa Rosada (Pink House) in Buenos Aires contains the offices of the president of Argentina and other government officials. The building also houses a presidential museum.

The National Congress building in Buenos Aires houses the two branches of the Argentine legislature. In front of the building stands the Monument to the Two Congresses.

Plaza de Mayo

The elegant Plaza de Mayo lies in the heart of Buenos Aires. Originally known as the Plaza de Armas, this site was chosen by Juan de Garay in 1580 for the first *cabildo* (town hall) of the new settlement of Buenos Aires. Spanish settlers had first established a settlement where Buenos Aires now stands in 1536, but they abandoned it only five years later because of Indian attacks. Juan de Garay and his fellow settlers reestablished the city and called it *Buenos Aires* (fair winds).

Today, the Plaza de Mayo is lined with restaurants, movie theaters, boutiques, bookstores, art galleries, and the city's finest hotels. At its east end stands the *Casa Rosada* (Pink House), which houses the office of the president of Argentina. To the west lies the Congress Building, which has its own square, known as the Plaza de Congreso. This square features a famous sculpture by Auguste Rodin, *The Thinker,* as well as the *zero kilometer stone,* from which all distances in Argentina are measured.

The barrios

Most of the northwestern, western, and southern sections of Buenos Aires consist of residential neighborhoods called *barrios,* each with its own churches, schools, and markets. Among the city's most colorful barrios is La Boca, known for its brightly painted houses and Italian restaurants.

Home to Buenos Aires' large Italian community, La Boca is also the birthplace of the tango, the first Latin American dance to become internationally popular. In addition, La Boca boasts one of Argentina's most famous soccer clubs, the Bocauniors.

The extreme differences between the rich and the poor in Buenos Aires are evident from one barrio to another. Many wealthy families live in mansions in the northern barrios or in elegant homes near the center of the city. In other barrios, thousands of poor families live in makeshift wooden shacks.

ECONOMY

Fertile farmland, Argentina's most important natural resource, traditionally was the basis of the country's economy. Agriculture and livestock production once supplied the nation with up to 95 percent of its export earnings. Today, however, service industries and manufacturing account for the majority of Argentina's economic activity.

Livestock production

Argentina is one of the world's most important producers of cattle and sheep, with about 4 percent of the world's cattle and about 3 percent of the world's sheep. After the Spanish explorer Pedro de Mendoza introduced cattle into Argentina in 1536, the animals were allowed to run wild, and they multiplied quickly on the lush grasses of the Pampas.

The development of refrigerator ships in the late 1800's helped turn these vast, wild herds of cattle into a profitable industry. With refrigeration, Argentina's cattle ranchers could ship meat to Europe and other markets without spoilage.

As the meat industry developed, foreign-owned meatpacking plants became more interested in the quality of the livestock they purchased, rather than just the weight of the animal. As a result, livestock producers in Argentina introduced higher-quality breeds into their stock and began using selective cross-breeding techniques.

Today, agriculture employs more than 10 percent of the country's workers, and the industry—including livestock and crop production—accounts for about 6 percent of Argentina's *gross domestic product* (GDP).

Agriculture

Wheat is a leading crop in Argentina. Huge fields of hard red winter wheat stretch across the fertile plain of the Pampas. Beef, corn, and soybeans are also major farm products. In dry areas throughout the country, sheep are raised, mainly for their wool. Other important products include apples, citrus fruits, grapes, milk, potatoes, poultry meat, sorghum, sugar cane, sunflower seeds, and tea.

While agriculture remains the foundation of the nation's economy, manufacturing and mining now provide many jobs.

Major mineral deposit
● *Petroleum*

Other mineral deposit
• *Lead*

Manufacturing center
• *Córdoba*

Commercial agriculture

Subsistence agriculture

Cereals and livestock

Grazing land

Chiefly forestland

Generally unproductive land

Petroleum
Iron ore, lead
Quebracho
Sugar cane
Cotton
Citrus fruit
Cattle *Córdoba*
Tungsten *Hogs* *Rosario*
Petroleum
Fruit *Corn*
Cattle
Buenos Aires
Natural gas, petroleum *Wheat*
Fruit
Iron ore
Sheep *Corn*
Natural gas, petroleum
Coal
Sheep

A gaucho and his herd present a classic image of the wild frontier, but most gauchos today work as ranch hands. Cattle raising and crop production formed the basis of Argentina's economy for many years.

Depositors lined up to remove their money from a Buenos Aires bank during the worst economic crisis in Argentina's history in 2001. The government limited bank withdrawals until 2003.

A busy lead-refining plant in the town of Comodoro Rivadavia reflects the increased economic activity in Patagonia. Oil and natural gas fields in Patagonia and the Piedmont supply much of Argentina's energy.

Fruit growing has developed rapidly in Argentina since the 1940's, particularly in the province of Mendoza, the center for the nation's vineyards.

Argentina's farms vary greatly in size. The owners of the huge, sprawling estates that cover much of the Pampas rent land to tenant farmers. They hire workers to tend livestock and help with planting and harvesting. In the north, many people own small farms, raising only enough to feed themselves. These farmers use horse-drawn equipment or rent machinery to help them work the land. Most large estates own modern equipment.

Manufacturing accounts for nearly one-fourth of Argentina's GDP. Manufacturing employs nearly 15 percent of all workers. Argentine factories include meat-packing plants and other food-processing facilities; leather-making factories; and plants that manufacture electrical equipment, printed materials, and textiles.

Most of the nation's factories are located in Buenos Aires and its suburbs. Factories in Córdoba manufacture automobiles, railroad cars, and tractors. The city is also a leading manufacturer of textiles and of glass and leather products. Rosario, a major inland seaport, has oil refineries as well as metal- and chemical-producing plants.

Service industries and manufacturing

Service industries account for about half of Argentina's GDP. These include government services, financial and insurance services, retail trade, tourism, and transportation. During the 1990's, retail trade experienced strong growth. In addition, international tourism has gained importance since the early 2000's.

A combine harvests wheat in Buenos Aires province. Wheat is one of the country's leading farm products, along with beef, corn, and soybeans.

HISTORY

When the Spanish explorer Juan Díaz de Solís landed on the shores of the Río de la Plata in 1516, he became the first European to reach what is now Argentina. Between 1527 and 1529, Sebastian Cabot also explored Argentina and founded the fort of Sancti Spiritus, the first Spanish settlement in the Río de la Plata Basin.

Jorge Luis Borges (1899–1986) Man of letters

Sancti Spiritus was destroyed by an Indian attack in 1529, but in 1536, Pedro de Mendoza established another settlement on the Río de la Plata where Buenos Aires now stands. Starvation and Indian raids killed many settlers, and the settlement was abandoned after five years.

Meanwhile, Spanish colonists living in what is now Peru crossed the Andes using the old Inca routes. They founded Santiago del Estero, Tucumán, and other towns in the northwest. In 1580, Juan de Garay founded a new settlement at Buenos Aires.

The Spaniards ruled what is now Argentina for 300 years, but they largely ignored the colony after they discovered it did not have huge deposits of gold and silver. Spanish rulers encouraged settlement in the area only to protect it from Portuguese expansion.

The Viceroyalty of La Plata

In 1776, the Spaniards created one large colony out of its territories in southeastern South America. The colony was called the Viceroyalty of La Plata. It consisted of what are now Argentina, Paraguay, Uruguay, and parts of Bolivia, Brazil, and Chile. Buenos Aires became the capital.

In 1806 and 1807, British troops tried to seize Buenos Aires, but the residents fought them off without help from Spain. This victory led the people of Buenos Aires to believe that they could fight off Spanish troops in a battle for independence.

TIMELINE

A.D. 1480	The Inca conquer the northwestern region of what is now Argentina.
1516	The first Spanish expedition arrives at the Rio de la Plata.
1527–1529	Sebastian Cabot explores the Rio de la Plata.
1536	The Spaniards found a short-lived settlement on the shores of the Rio de la Plata.
1580	Juan de Garay establishes permanent settlement of Buenos Aires.
1776	Viceroyalty of the Río de la Plata is created.
1806 and 1807	British troops invade Buenos Aires.
1810	Buenos Aires forms an independent government.
1816	Argentine provinces declare their independence from Spain.
1829–1852	Manuel de Rosas rules the United Provinces of the Río de la Plata as dictator.
1853	All the Argentine provinces except Buenos Aires agree to adopt a federal constitution.
1860	The country takes the name of Argentina.
1862	Buenos Aires joins Argentina.
1877	Refrigerated meat is first exported from Argentina.
1881	Patagonia becomes part of Argentina.
1912	The Sáenz Peña Law reforms national elections.
1929	The Great Depression begins.
1930	Army officers overthrow the elected government.
1943	Juan Perón begins his rise to power.
1946	Perón is elected president. His second wife, Eva, serves as his chief assistant until her death in 1952.
1955	A military revolt overthrows the Perón dictatorship. Perón flees the country.
1973	Perón returns from exile and is elected president.
1974	Perón dies, and his third wife, Isabel, becomes president.
1976	Military leaders remove Isabel Perón from office.
1982	Argentina loses a war with the United Kingdom over control of the Falkland Islands.
1983	Civilian rule is restored following free elections.
1985–1986	Senior military officers are convicted of terrorism, including murders and torture, and are sentenced to prison.
Late 1990's	Country is in a recession.
2000–2002	Government increases taxes and makes massive spending cuts.
2008	An export tax increase on agriculture products triggers nationwide strikes by farmers until the tax is canceled.

Huge crowds gather in the Plaza de Mayo in Buenos Aires in support of Juan Perón. Perón won the support of Argentina's working classes by giving them higher wages, more paid holidays, and other benefits.

➡

Argentinians rallied to support farmers in 2008, protesting export taxes on agricultural products. Farmers went on strike throughout the country, demanding that the tax be dropped.

Juan Perón,
(1895–1974)
President, 1946-1955,
1973-1974

Eva Perón
(1919-1952)
First lady of Argentina,
1946-1952

In 1810, Buenos Aires set up an independent government to administer the Viceroyalty of La Plata. However, the provinces outside Argentina opposed the movement and eventually broke away.

Beginning in 1812, José de San Martín, an Argentine general, led the fight against Spanish rule. In 1816, representatives of the Argentine provinces formally declared their independence at the Congress of Tucumán. The new country took the name the United Provinces of the Río de la Plata.

Two new constitutions

In 1852, delegates from all provinces except Buenos Aires drew up a constitution that set up a strong national government. In 1860, the country took the name of Argentina, and in 1862, Buenos Aires agreed to the union and became the nation's capital. In 1881, Argentina annexed the territory of Patagonia.

The people of Argentina enjoyed nearly 70 years of political stability. But Argentina was hit hard by the Great Depression, a worldwide economic slump that began in 1929. Military leaders took over the government, and Colonel Juan Perón was elected president in 1946. In 1955, the army and navy revolted, and Perón fled the country.

During the late 1960's and early 1970's, a succession of military and civilian governments ruled Argentina. The country's economic problems worsened. In 1973, Perón returned from exile to be president again. His third wife, Isabel, was elected vice president and became president when he died in 1974.

In 1976, the military seized power and launched a period of violent repression. Thousands were imprisoned without trial, tortured, and killed. In 1982, Argentina seized the British-controlled Falkland Islands. The United Kingdom won an ensuing 74-day war.

Unrest forced the military to relinquish power. In 1983, Raúl Alfonsín was elected president. In 1989, amid worsening inflation, Carlos Saúl Menem was elected leader. He introduced an emergency economic program, and the rate of inflation dropped.

In 1994, a new constitution took effect, allowing a president to serve two four-year terms instead of one six-year term. Problems in 2001, however, resulted in a period in which three leaders held the office of president for only a matter of hours or days before Eduardo Duhalde became leader in 2002. He was replaced by Néstor Kirchner in 2003.

In 2007, Argentine first lady Cristina Fernández de Kirchner was elected president. She promised to continue her husband's left-leaning policies. She was re-elected to a second term in 2011. Néstor Kirchner died in 2010.

ARMENIA

Armenia, once a republic of the Soviet Union, is an independent country and a member of the Commonwealth of Independent States (CIS). Armenia was under the strict control of the Soviet central government until 1990, when the republic called for greater control of its own affairs. In the midst of political upheaval in the Soviet Union in August 1991, Armenia declared its independence. When the Soviet Union was dissolved in December 1991, Armenia joined the newly established CIS.

Earthquakes occur frequently in the geologically active region. A violent quake devastated western Armenia in 1988, killing about 25,000 people and causing severe property damage. The earthquake virtually destroyed the town of Spitak, which lay at its epicenter.

Manufacturing and mining account for much of Armenia's economic production. The chief industries process diamonds and make chemicals, clothing, electronic products, machinery, and processed foods. Armenia is a leading distiller of cognac. Armenian mines produce copper, gold, lead, and zinc. Construction is also a major economic activity. About half of the country's workers are employed in agriculture. Farmers produce such crops as barley, potatoes, tomatoes, wheat, and wine grapes.

The small, ancient land of Armenia has seen periods of great power and prosperity, but there have also been times in which its people suffered greatly under foreign rule. From earliest times, the Armenian people have had to fight hard to preserve their ethnic culture.

King Tigran II, who came to power in 95 B.C., built an independent Armenian empire that stretched from the Caspian Sea to the Mediterranean Sea. The Romans defeated Tigran in 55 B.C. and made Armenia part of the Roman Empire.

The Arabs conquered the region in the A.D. 600's, followed by the Seljuks in the mid-1000's. In 1375, the Armenian kingdom fell to Mamluks invaders, and by 1514, the Ottomans had gained control of Armenia. The Persians took over eastern Armenia in 1639 and ruled the region until 1828, when Russia annexed it. Western Armenia remained under Ottoman control.

FACTS

Official name:	Haikaken Hanrapetoutioun (Republic of Armenia)
Capital:	Yerevan
Terrain:	Armenian Highland with mountains; little forestland; fast flowing rivers; good soil in Aras River valley
Area:	11,484 mi² (29,743 km²)
Climate:	Highland continental, hot summers, cold winters
Main rivers:	Hrazdan, a tributary of the Aras
Highest elevation:	Mount Aragats, 13,419 ft (4,090 m)
Lowest elevation:	Aras River, 1,475 ft (450 m) below sea level
Form of government:	Republic
Head of state:	President
Head of government:	Prime minister
Administrative areas:	11 marzer (provinces)
Legislature:	Azgayin Zhoghov (National Assembly) with 131 members serving four-year terms
Court system:	Constitutional Court; Court of Cassation (Appeals Court)
Armed forces:	42,100 troops
National holiday:	Independence Day - September 21 (1991)
Estimated 2010 population:	2,983,000
Population density:	260 persons per mi² (100 per km²)
Population distribution:	64% urban, 36% rural
Life expectancy in years:	Male, 69; female, 76
Doctors per 1,000 people:	3.7
Birth rate per 1,000:	13
Death rate per 1,000:	9
Infant mortality:	22 deaths per 1,000 live births
Age structure:	0-14: 19%; 15-64: 70%; 65 and over: 11%
Internet users per 100 people:	6
Internet code:	.am
Languages spoken:	Armenian, Yezidi, Russian
Religions:	Armenian Apostolic 94.7%, other Christian 4%, Yezidi 1.3%
Currency:	Dram
Gross domestic product (GDP) in 2008:	$11.97 billion U.S.
Real annual growth rate (2008):	7.6%
GDP per capita (2008):	$3,999 U.S.
Goods exported:	Copper, diamonds, food, iron
Goods imported:	Diamonds, food, machinery, natural gas, petroleum, tobacco products
Trading partners:	Belgium, Germany, Iran, Russia, United States

Armenia lies on the Armenian Plateau, a rugged highland. The average altitude of Armenia is 5,000 ft. (1,500 m).

The Jermuk Waterfall spills down the rocky mountainside in a spectacular sight in the Armenian highlands. Thousands of people from neighboring regions come to the Jermuk Spa to enjoy the pure mountain air and picturesque scenery.

During World War I (1914–1918), Armenia became a battleground between the Ottoman Empire and Russia. The Ottomans deported countless Armenians to what is now Syria to keep them from aiding Russia. About 1-1/2 million Armenians died of starvation or were killed by Ottoman soldiers, Arabs, and Kurds. The survivors fled to Russian Armenia, where they established an independent Armenian republic in 1918.

When conflicts resurfaced between the Armenian republic and the Ottoman Empire, the Armenians reluctantly turned to Soviet Russia for protection. Eastern Armenia became a Soviet republic, while the Ottomans kept the rest of Armenia. In 1922, Soviet Armenia became part of the Soviet Union.

Most of the people are Armenians, but some Azerbaijanis, Kurds, and Russians also live in Armenia. In the late 1980's and early 1990's, disputes arose over the status of Nagorno-Karabakh—a district in neighboring Azerbaijan where most of the people are Armenian. The disputes led to fighting between Armenians and Azerbaijanis. By late 1993, Armenia controlled the district and occupied a small surrounding area of Azerbaijan. In 1994, the two countries agreed to a cease-fire. However, the conflict over the territory remained unresolved.

In 1996, many Armenians protested that the presidential election held that year was marred by fraud. In 1998, the president resigned and the prime minister was elected as the new president. In October 1999, gunmen entered the parliament building and assassinated Armenia's prime minister and several other officials. The gunmen were caught, and replacements were appointed for the slain officials.

According to observers, all of Armenia's presidential elections from 1995 through 2008 have been flawed. Following the 2008 election, mass protests broke out between demonstrators and security forces, and the leading opposition group was banned. The ban was lifted in 2011.

AUSTRALIA

The name Australia comes from the Latin word *australis*, meaning southern. And because it lies entirely within the Southern Hemisphere, Australia is often referred to as "down under." Australia is the only country that is also a continent. In area, it ranks as the sixth largest country and the smallest continent.

Australia is a dry, thinly populated land. Only a few coastal areas receive enough rainfall to support a large population. Most of Australia's people live in the south-eastern coastal region, which includes the country's two largest cities—Sydney and Melbourne. Australia's vast interior is mostly desert or dry grassland and has few settlements. The country as a whole has an average of only 7 persons per square mile (3 per square kilometer).

Australia is famous for its vast open spaces, bright sunshine, enormous numbers of sheep and cattle, and unusual wildlife. Kangaroos, koalas, platypuses, and wombats are only a few of the many exotic animals that live there.

Along the northeast coast of the continent lies the Great Barrier Reef, the largest group of coral reefs in the world. This unique area supports an unmatched variety and quantity of coral polyps.

The first Australians were a people known today as Aborigines. The Aborigines had lived in Australia for about 50,000 years before the first European settlers arrived. Since then, the number of Europeans has steadily increased and the number of Aborigines has dramatically declined. Today, the vast majority of Australians have European or mixed European ancestry.

Australia is one of the world's developed countries, with bustling cities, modern factories, and highly productive farms and mines. Its economy is increasingly diversified, with service industries dominating and a gradual shift of the value of exports from mining to manufacturing. The income from the nation's exports has given most of Australia's people a high standard of living.

Great Britain settled Australia as a prison colony in the late 1700's, and most Australian people are of British ancestry. The early settlers brought many British customs that remain part of Australian culture today. For example, tea is still a favorite hot drink, and people drive on the left side of the road, as the British do. In addition, English—with many British terms—is the official language. Nevertheless, just as the people of Australia have given the language a distinctive slant with their own terms and pronunciation, so have they developed a way of life all their own.

AUSTRALIA TODAY

Australia has a population of more than 21 million. Approximately 80 percent of the people live in the southeastern quarter of the country, with most of the remaining population living along the northeast and extreme southwest coasts. Canberra, the national capital and the largest inland city, lies about 80 miles (130 kilometers) from the ocean. Most of Australia's people live in cities and towns, making it one of the world's most urbanized countries. About 75 percent of the population live in cities of more than 100,000 people. Only about 12 percent of Australia's people live in rural areas.

The Commonwealth of Australia is a federation of six states—New South Wales, Queensland, South Australia, Tasmania, Victoria, and Western Australia. Each state has its own government. Australia's Constitution gives certain powers to the federal government and leaves all others to the states. Australia also has two mainland territories—the Australian Capital Territory and the Northern Territory—that do not have the status of statehood. Eight additional territories lie outside of the mainland.

In Australia's parliamentary system of government, the national government is controlled by the political party or combination of parties with a majority of seats in the lower house of Parliament. The leader of the majority heads the government as prime minister. The prime minister appoints members of Parliament to serve as ministers, the heads of government departments. The prime minister and department heads form the Cabinet, which establishes major government policies.

Australia is a constitutional monarchy like the United Kingdom. The British monarch is also Australia's monarch and head of state. However, the monarch serves mainly as a symbol of the historical ties between the two countries and has little or no power in the Australian government. Australia is a member of the Commonwealth of Nations, with many other former British colonies. In the late 1990's, Australia considered becoming a republic, but the idea was ultimately voted down.

Australia is one of the world's rich, developed countries. Most developed countries have become rich through the export of manufactured goods, but Australia's wealth has come chiefly from farming and mining.

FACTS

Official name:	Commonwealth of Australia
Capital:	Canberra
Terrain:	Mostly low plateau with deserts; fertile plain in southeast
Area:	2,969,907 mi² (7,692,024 km²)
Climate:	Generally arid to semiarid; temperate in south and east; tropical in north
Main rivers:	Murray, Darling, Lachlan, Murrumbidgee
Highest elevation:	Mount Kosciuszko, 7,310 ft (2,228 m)
Lowest elevation:	Lake Eyre, 52 ft (16 m) below sea level
Form of government:	Constitutional monarchy; in practice, a parliamentary democracy
Head of state:	British monarch, represented by governor general
Head of government:	Prime minister
Administrative areas:	6 states, 2 territories
Legislature:	Federal Parliament consisting of the Senate with 76 members serving six-year terms and the House of Representatives with 150 members serving three-year terms
Court system:	High Court
Armed forces:	54,700 troops
National holiday:	Australia Day - January 26 (1788); ANZAC Day - April 25 (1915)
Estimated 2010 population:	21,865,000
Population density:	7 persons per mi² (3 per km²)
Population distribution:	88% urban, 12% rural
Life expectancy in years:	Male, 79; female, 84
Doctors per 1,000 people:	2.5
Birth rate per 1,000:	14
Death rate per 1,000:	7
Infant mortality:	5 deaths per 1,000 live births
Age structure:	0-14: 19%; 15-64: 68%; 65 and over: 13%
Internet users per 100 people:	72
Internet code:	.au
Languages spoken:	English, Chinese, Italian, Greek
Religions:	Roman Catholic 25.8%, Anglican 18.7%, other Christian 19.3%, Buddhist 2.1%, other 34.1%
Currency:	Australian dollar
Gross domestic product (GDP) in 2008:	$1.032 trillion U.S.
Real annual growth rate (2008):	2.2%
GDP per capita (2008):	$49,175 U.S.
Goods exported:	Alumina, beef, coal, iron ore, petroleum products, wheat, wool
Goods imported:	Electrical appliances, industrial machinery, office equipment, petroleum products, telecommunications equipment
Trading partners:	China, Germany, Japan, New Zealand, United Kingdom, United States

North

Australia, the world's sixth largest country, is the only country that is also a continent. Its territory includes the island of Tasmania, which lies off its southern coast.

AUSTRALIA

0	250		500 Miles
0	250	500	750 Kilometers

Like many other developed countries, Australia faces the problems of continuing inflation and unemployment, as well as a growing International debt. The country's hopes for economic growth are closely tied to the growth of its mining industry.

Uranium is Australia's most valuable undeveloped mineral resource. However, Australians who oppose nuclear power because of its potential hazards also oppose plans to mine and export uranium. In addition, some of the richest uranium deposits lie in the traditional tribal lands of the Aborigines. Aboriginal groups are trying to obtain legal control over their tribal lands, including the uranium-mining areas.

ENVIRONMENT

The northern third of the Australian continent lies in the tropics and is warm or hot the year around. The rest of the country lies south of the tropics and has warm summers and mild or cool winters. Rainfall is seasonal in most of Australia.

Wet and dry seasons

During the wet season, heavy downpours and violent storms cause floods in many parts of Australia. However, the droughts that plague the nation are usually a far more serious problem. Nearly every section of Australia has a drought during the annual dry season. These droughts can cause severe water shortages that require strict conservation measures. In addition, destructive bush fires are more common during droughts.

Australia's rivers are one of its most vital resources. They provide the towns and cities with drinking water and supply farmers with much-needed water for irrigation. However, most of Australia's rivers are dry at least part of the year, so dams and reservoirs on all the largest rivers store water for use during the dry season.

Land regions

Australia can be divided into three major land regions. The easternmost region is the Eastern Highlands—sometimes known as the Great Dividing Range because the mountains divide the flow of rivers in the region. The Eastern Highlands include the highest elevations in Australia. A low plain bordered by sandy beaches and rocky cliffs stretches along the Pacific coast. More rain falls on this coastal plain than anywhere else in the country.

The highlands consist mainly of high plateaus broken in many places by gorges, hills, and low mountain ranges. Grass or forests cover some plateaus in the Eastern Highlands, but many plateaus have fertile soils and are used as cropland. The southeastern section of the plain, from Brisbane to Melbourne, is by far the most heavily populated part of Australia. The Australian Alps, with their Snowy Mountains, lie in the southern part of this region. The Murray River, Australia's longest permanently flowing river, starts in the Snowy Mountains.

Australia's land regions consist of the Western Plateau, which covers much of the continent and contains three deserts; the Central Lowlands, which have the lowest elevations in Australia; and the Eastern Highlands, which run down the Pacific coast.

The Central Lowlands, Australia's second major region, is a generally flat area with infrequent rainfall, except along the north and south coasts and near the Eastern Highlands. Farmers in the southern part of the Lowlands grow wheat, but most of the region is too dry or too hot for crops. However, the coarse grass or shrubs that cover much of the land make it suitable for grazing livestock. The two largest towns in the region have fewer than 30,000 people each.

The Western Plateau, Australia's third major region, covers the western two-thirds of Australia. A vast, dry, treeless plateau extends about 400 miles (640 kilometers) along the region's southern edge, while deserts stretch across the central part. Most of the desert area consists of swirling sands that often drift into giant dunes. Where the deserts give way to land covered by grass and shrubs, the land can be used to graze livestock. The extreme north and southwest have the region's heaviest rainfall, and most of its cropland. Adelaide and Perth are the region's two largest cities.

The Flinders Ranges of southern Australia extends north from Gulf St. Vincent, an arm of the Indian Ocean that cuts into the land at Adelaide. Wind and water have eroded the reddish rocks of the range, giving them sharp outlines.

Salt crystals cover the dry bed of Lake Eyre in southern Australia. Most of Australia's natural lakes are dry for months or years at a time. Playas, dry beds of salt or clay that fill with water only after heavy rains, are common in South Australia and Western Australia.

WILDLIFE

At one time, all the world's continents were part of one huge land mass. The region that is now Australia became separated from this land mass about 200 million years ago, and as a result, the animals of Australia developed differently from those of other continents. For example, marsupials—animals with pouches—are Australia's most unusual and famous creatures.

Unique animal life

Most mammals give birth to relatively well developed offspring, but Australia's marsupials give birth to tiny, poorly developed offspring. The marsupial newborn undergoes most of its development while attached to one of its mother's nipples in a pouch called the marsupium. Kangaroos, koalas, wallabies, and wombats are marsupials. Australia has about 200 species of marsupials, all of which have pouches.

The platypus and the echidna are the only mammals in the world that hatch their young from eggs. Platypuses, found only in Australia, live in long burrows that they dig in the banks of streams. Except for females with their young, each platypus lives in its own burrow. Platypuses can walk on land as well as swim, and they use their bills to scoop their food from the bottom of streams. Echidnas live in New Guinea as well as in Australia.

Australia also has several hundred species of native birds. They include the world's only black swans and about 60 kinds of cockatoos, parakeets, and other parrots. The emu, a large, flightless bird, has long legs for running and stands 5-1/2 feet (1.7 meters) high. The kookaburra, one of Australia's best-known birds, nests in tree holes and has a call that sounds like a loud laugh.

Plant life

Two kinds of native plants dominate Australia's landscape. Varieties of acacias and eucalyptuses are the most common shrubs in dry areas and the most common trees in moist areas. Acacias, which

Impressive jumpers, kangaroos can reach a top speed of about 30 miles (48 kilometers) per hour. Each jump covers an average distance of about 13 feet (4 meters).

A female koala and her young rest in a forked tree. The koala, a native of Australia's eastern forests, was near extinction in the 1920's. Today, as a protected species, this charming creature has been making a slow comeback.

Two species of crocodile live in the coastal regions of tropical north and east Australia. The saltwater crocodile can be ferocious, while the smaller freshwater species is considered harmless.

Kangaroos are the symbol of Australia to people throughout the world. They belong to a family of about 60 species of mammals. Kangaroos range in size from the huge red kangaroo, which can stand up to 6 feet (1.8 meters) tall, to a small tree kangaroo measuring only about 2 feet (0.6 meter) long.

Emus cannot fly, but they can run swiftly. They are found everywhere in the country except the rain forests.

Australians call wattles, bear their seeds in pods. Australia has about 700 species of acacias, many with brightly colored flowers.

Eucalyptuses, or eucalypts, as the Australians call them, are the most widespread plants in the country. Most species have narrow, leathery leaves that contain a fragrant oil. Scrubby eucalyptuses cover large areas of Australia's hot, dry interior. Eucalyptus trees, known as gum trees or gums, are the tallest trees in the country and among the tallest in the world. Some types of eucalyptus may grow to a height of 330 feet (100 meters). At one time, the eucalyptus grew only in Australia and on a few islands to the north. But these trees have been planted in California, Hawaii, and other warm areas.

Australia has thousands of wildflowers. Many are desert species whose seeds lie buried in the ground until heavy rain brings them to life. These plants can make deserts look like gardens right after a desert rain. The waratah, a tall shrub found only in Australia, grows under trees in open forests and bears large, bright-red flowers. Its name is an Aboriginal word meaning red-flowering tree.

To a great extent, Australia's wildlife has suffered at the hands of the European settlers, who have greatly changed the environment since they arrived in the 1700's. Many wildlife species became extinct in the 1800's, and many others have been lost since then. Today, more than 1,000 Australian plants and animals are threatened. Many others are protected.

Galah parrots are pink and gray cockatoos that live in many inland areas of Australia. Many people keep these parrots as pets. Galahs eat the seeds of grasses and other plants.

THE ABORIGINES

Aborigines are Australians whose ancestors were the first people to live in Australia. The name comes from the Latin phrase *ab origine*, meaning from the beginning. Most scientists believe the ancestors of today's Aborigines arrived in Australia about 50,000 years ago from Southeast Asia. By the time the Europeans arrived, in 1788, 250 tribes of Aborigines had developed, each with its own language.

Young Aborigines are born with light-brown or blond hair that gradually darkens during childhood.

Traditional culture

The Aborigines traditionally lived by hunting and gathering food. They did not settle in one place but roamed over limited areas of the countryside. They made weapons, tools, and utensils from local resources. For men, the most important weapons were spears, which they used in fighting, hunting, and fishing. Women gathered vegetables and small animals.

Australia's first people lived close to nature. They knew the habits of all the creatures and plants around them. In addition, all adult Aborigines knew where they could find water within their territory. Little girls went out gathering food with their mothers and other women. Boys began to practice throwing toy spears quite early.

The size of a tribe depended partly on the amount of food and water in its territory. A tribe had no political chief or formal government, but older, respected men generally made important tribal decisions and directed the ceremonies. Each tribe consisted of various subgroups. Based on their ties to a common ancestor, these family groups owned certain lands and conducted ceremonial rituals.

Religion linked the Aborigines to the land and nature through ancestral beings who, according to Aboriginal beliefs, had created the world long ago in a time called the Dreaming, or Dreamtime. These beings never died, but merged with nature to live in sacred beliefs and rituals. Aborigines could renew their ties with the Dreaming through their rituals.

Spears were the Aborigines' most important weapons, and they were used for fighting, hunting, and fishing. A tool called a spearthrower increased the spear's speed and force. The men also used boomerangs for hunting.

Ceremonies called corroborees consist of songs and dances that are performed for amusement and relaxation rather than for religious reasons. The ancient dances are passed on to the children unchanged.

Thousands of people gathered in 2008 in Melbourne and other Australian cities to hear Prime Minister Kevin Rudd deliver a historic apology to the Aboriginal people for injustices committed over 200 years of white settlement.

Cool, dark caves inside Uluru, or Ayers Rock, have sheltered Aborigines for thousands of years. The land where Uluru stands was returned to its traditional Aborigine owners in 1985.

1700's to today

At first, many Aborigines supposed that the European newcomers, with their pale skins, were the spirits of their own dead relatives. But the Aborigines' image of Europeans soon changed to that of evil spirits. The Europeans killed many Aborigines or forced them from their homes. Other Aborigines died from diseases, such as smallpox and measles, brought by the newcomers.

Today, Australia has only about 420,000 Aborigines—about 2 percent of the country's population. Some Aborigines live partly as their ancestors did, with their own customs, beliefs, and language. Many others lost most or all of their traditions but were excluded from mainstream society.

In general, the Aborigines lag far behind white Australians in both education and income. In addition, most lack decent housing and proper health care.

Only since the 1930's has the Australian government worked to include the Aborigines more closely in the country's economic and political life. In 1967, the Constitution was changed so that Aborigines were allowed to vote and receive social service benefits.

Today, Aborigines have made progress in gaining land rights and overcoming discrimination. However, they still face difficulties in such issues as health, education, and employment.

THE OUTBACK

Australians call the countryside the bush. The term outback refers specifically to the interior of the country, which consists mainly of open countryside, including vast expanses of grazing land. Only about 12 percent of Australia's people live in these rural areas. Many live extremely isolated lives on cattle or sheep ranches, called stations. The largest stations cover about 4,000 square miles (10,000 square kilometers) and may be 100 miles (160 kilometers) or more from the nearest town.

The outback has few paved roads, so travel by automobile is difficult or impossible. Floods sometimes close roads for weeks at a time. Most wealthy farm families own a light airplane, which they use for trips to town. Other families may get to town only a few times a year, making it difficult to maintain supplies of food and other necessities.

Many children in remote areas of the outback receive elementary and secondary education at home by connecting to so-called "schools of the air." They can connect to such schools through two-way radio, television, fax, and the Internet. A movement

Musgrave Ranges, an Aboriginal reserve in the heart of the Australian continent, is typical of much of the outback landscape. Its natural vegetation is sparse grassland with scrubby trees and bushes.

The population distribution of Australia is uneven. Most people live in the big cities along the southern and eastern coasts of the continent. The southwest corner of Australia also has a sizable population, but the dry center of the country has few settlements. The Aboriginal people live in small groups, mainly in rural areas.

The rodeo allows farmhands in the outback to put aside their daily chores and show off their riding and roping skills. Riders compete in events such as bareback bronc riding, calf roping, and steer wrestling.

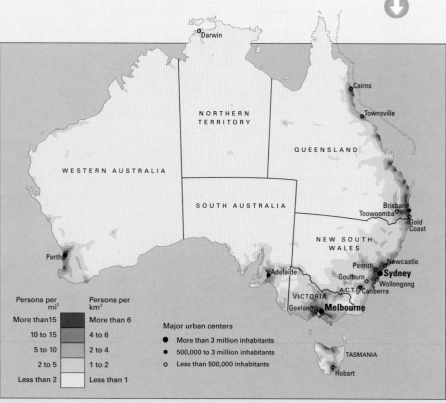

Darwin

NORTHERN TERRITORY

Cairns

Townsville

QUEENSLAND

WESTERN AUSTRALIA

SOUTH AUSTRALIA

Brisbane
Toowoomba
Gold Coast

NEW SOUTH WALES

Perth

Penrith
Newcastle
Adelaide
Goulburn
Sydney
A.C.T.
Wollongong
Canberra

VICTORIA
Geelong
Melbourne

Persons per mi²	Persons per km²
More than 15	More than 6
10 to 15	4 to 6
5 to 10	2 to 4
2 to 5	1 to 2
Less than 2	Less than 1

Major urban centers
- More than 3 million inhabitants
- 500,000 to 3 million inhabitants
- Less than 500,000 inhabitants

TASMANIA

Hobart

A wind pump draws water from the Great Artesian Basin in central Australia. This vast underground rock formation extends across much of eastern Australia. The water it provides is often too salty for people to drink, but is suitable for livestock

Farmers on horseback check on their beef cattle in a remote part of the outback.

is underway to provide students in the outback with computers, so that they can attend virtual classes by satellite.

The largest settlements in rural Australia are the widely scattered towns that developed to support Australia's mining industry. For example, Kalgoorlie, a town in Western Australia, is the center of Australia's major gold and nickel fields. The town lies in an arid region about 370 miles (600 kilometers) east of Perth. Water must be pumped from about 350 miles (560 kilometers) away, and the town must supply most of its own needs.

All rural areas in Australia are subject to such disasters as droughts, floods, and bushfires. Because they share the threat of frequent catastrophe, rural Australians tend to develop strong ties with one another. Many communities have their own traditional fairs, festivals, and sports competitions.

Outback animal life is varied and includes native species such as kangaroos, wallabies, and emus, and introduced species, such as rabbits. The size of agricultural stations makes it prohibitive to fence off crops from the wildlife, and some of these animals have caused extensive damage to crop and grazing lands. Wild rabbits, in particular, have been destructive. In recent years, rabbits have been the target of a drive to wipe them out with a deliberately introduced disease, myxomatosis.

The standard of living in rural Australia is generally lower than it is in the cities. Still, most farm families own their farms and live in comfortable wood or brick houses with electric power. A growing number have air conditioning. Economic difficulties have come as a result of a drop in demand and prices for many farm products, especially wheat.

The Royal Flying Doctor Service brings emergency medical care to rural and remote areas of Australia through airplanes equipped with modern medical equipment.

AGRICULTURE

Much of Australia's wealth comes from the farmland that covers about 60 percent of the nation. However, most of this land is dry grazing land. Crops are grown on only about 10 percent of the farmland, but farmers use modern agricultural methods that make the cropland highly productive.

Farms on the east coast of Queensland grow bananas, pineapples, sugar cane, and other crops that need a wet, tropical climate. Wine grapes and oranges grow in some parts of the country, and apples and pears grow in all the states. Cattle and calves, wheat, and wool are Australia's leading farm products and are also the country's chief agricultural exports. Australia ranks as the world's largest wool producer and exporter.

A flock of Merino sheep await a shearing on an Australian sheep station.

Sheep, cattle, and wheat

Sheep and cattle are raised in all Australian states, though some states raise far more than others. New South Wales and Western Australia together raise more than half of the country's sheep and produce about half of its wool. Most farmers also raise cattle and grow wheat.

New South Wales and Queensland raise more than half of Australia's beef cattle. The mild Australian winters allow beef cattle to graze throughout the year. Australian farmers can therefore produce beef at lower costs than farmers in most of Europe and the United States, where cattle must be housed and hand-fed in winter.

Australia's wheat production grew rapidly during the 1980's. Today, Australia is among the world's leading wheat producers. Western Australia and New South Wales produce the most wheat. Australian white wheats are world renowned for their resistance to disease.

A harvester picks grapes on the Southern Highlands of New South Wales. Australian vineyards produce many outstanding vintages of wine that are consumed throughout the world

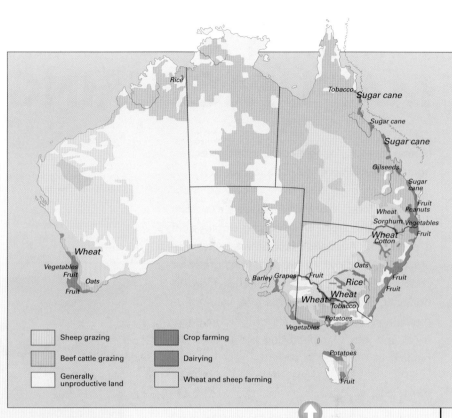

Wheat production is heavily concentrated in New South Wales and Western Australia. The crop is grown in areas where there is medium rainfall and moderate temperatures. Australia sells large quantities of wheat to Indonesia, Japan, and other Asian countries.

Most of Australia's cropland is concentrated along the country's southwest, southeast, and east coasts, the only areas that get enough rainfall for growing crops. The drier inland areas support sheep and cattle grazing and wheat growing.

Modern farm methods

At one time, farmers believed that their production depended entirely on how much land they used. Today, farmers know that they can greatly increase production without increasing the area they cultivate. For example, by adding small quantities of certain elements to the land, scientists have transformed areas of poor land, such as the Ninety Mile Desert in South Australia, into good grazing land. Research workers have also devoted a great deal of effort to developing new, improved varieties of rice, sugar, and wheat.

Because the amount of Australia's rainfall is low and unpredictable, the full development of Australia's agricultural resources also depends on irrigation. Irrigation gives farmers control over the water supply and allows them to grow fruits and vegetables in areas where it would not otherwise be possible. Irrigation also enables farmers to obtain higher crop yields.

Workers bring in the tomato harvest in the Northern Territory. Most of the crop farming in the Northern Territory takes place in the northern and south-central regions.

MINING AND MANUFACTURING

Mining has long been a cornerstone of economic development in Australia. However, manufacturing has also come to play an important part. Today, many of Australia's exports are mined products. While mining employs only 1 percent of Australia's labor force today, manufacturing employs 11 percent. Mining also accounts for only 5 percent of Australia's gross domestic product (GDP)—the total value of goods and services produced within the country in one year. Manufacturing accounts for 12 percent of the GDP.

Mineral wealth

Australia has become one of the world's major mining countries, ranking first in the production of bauxite, diamonds, lead, and zircon. The nation is also a leading producer of coal, copper, gold, iron ore, manganese, nickel, silver, tin, titanium, and zinc. Nearly all the world's high-quality opals are mined in Australia.

The continent's energy-producing resources are among the richest in the world. Australia has enough coal and natural gas to meet all its energy needs. It has large reserves of brown and black coal, which are used to produce Australia's electricity. Australia is a leading exporter of coal and natural gas. However, Australia needs to import oil.

Many of Australia's mineral deposits lie in the country's dry areas, far from major settlements. Such deposits are extremely expensive to mine. Roads and railroads to the mining sites must be constructed, and towns must be built for the miners and their families. The costs of mining development in Australia are so high that the mining industry depends heavily on support from foreign investors.

Two men pan for gold in the Hill End district of New South Wales. During the 1850's, the rich gold fields of New South Wales and Victoria attracted adventurers from all over the world who hoped to make quick fortunes.

Australia's abundance of deposits makes it one of the world's major mining countries. The country is almost self-sufficient in minerals but relies heavily on foreign investors to pay the costs of mining.

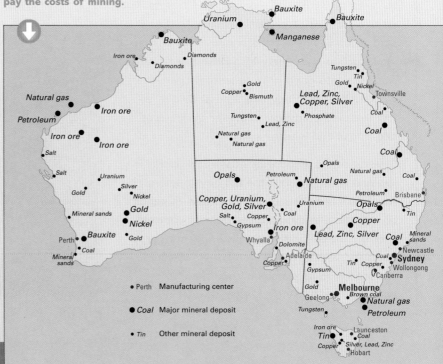

A mountain of salt piles up at Port Hedland in Western Australia. The country produces about 11 million short tons (10 million metric tons) of salt per year. Australia exports much of its salt production, mainly to Japan.

Manufacturing power

After World War II (1939–1945), the Australian government determined that its industry should become self-sufficient, and that at least one brand of automobile must be made entirely in Australia—including the engine and parts. Australia today produces most of its own metals, such as iron and steel, for local industrial use.

But unlike most other developed countries, Australia still imports more manufactured goods than it exports. Australian factories produce most of the nation's consumer goods, such as processed foods and household articles. But many producer goods, such as factory machinery and construction equipment, must be imported. Iron and steel are the chief exceptions.

Most of Australia's factories specialize in assembly work and light manufacturing, and many plants process farm products or minerals for export. Automobiles, chemicals, metals, paper, and processed foods are among Australia's leading products. New South Wales and Victoria are the chief manufacturing states, with most factories located in and around Sydney and Melbourne.

Opals only show their real beauty after polishing. These valuable gems are mined in South Australia. The country produces nearly all of the world's high-quality opals.

The Cadia mine is a large open cut gold and copper mine located near Orange in New South Wales. The mine was opened in 1998. Australia is a leading producer of gold and copper as well as many other minerals.

SYDNEY AND MELBOURNE

The two largest cities in Australia are Sydney and Melbourne. Sydney has more than 4 million people, and Melbourne, about 3-1/2 million. Both are state capitals and the major commercial, industrial, and cultural centers of their states. Each city was laid out near the mouth of a river and close to a good ocean harbor. The rivers provided drinking water, and the harbors enabled the settlements to develop into centers of trade and immigration.

Today, both Sydney and Melbourne have problems common to big cities everywhere. Poor inner-city areas have high rates of unemployment and crime. Air pollution and rush-hour traffic jams plague both cities' residents.

Sydney

The oldest and largest city in Australia, Sydney is the capital of the state of New South Wales. The city and its suburbs cover about 4,690 square miles (12,100 square kilometers) on the country's southeastern coast. The British founded Sydney as a prison colony in 1788.

The city's mild climate enables the people of Sydney, called Sydneysiders, to enjoy the outdoors during most of the year. Sydney's vast harbor makes the area famous for water sports, such as sailing. Most Sydneysiders have British ancestors, but many other Europeans and smaller numbers of Asians have settled in the city since the mid-1900's. Sydney also has a few thousand Aborigines. Most Sydney families own a house and garden in one of the suburbs. Sydney has almost no slums, but some of the Aborigines live in sub-standard housing.

High-rise buildings dominate the skyline of Sydney's central business district. Sydney Harbour, one of the world's major ports, is also Australia's most important port for shipping farm products. The city has thousands of manufacturing

Sydney Harbour is one of the world's major ports. Its white shell-like Opera House, and the Sydney Harbour Bridge are landmarks recognized around the world. The bridge links the city center with suburbs on the north shore of the harbor.

Bondi Beach is a beautiful stretch of sand close to the center of bustling downtown Sydney. The beach provides a peaceful refuge for many Sydneysiders during the warm weather season.

plants. Leading factory products include machinery equipment, chemical and paper goods, metal products, and food products. Sydney also serves as the banking and business center of New South Wales.

Sydney's famous Opera House, completed in 1973, includes facilities for concerts, opera, ballet, and theater. Many architects consider it one of the finest buildings constructed during the 1900's.

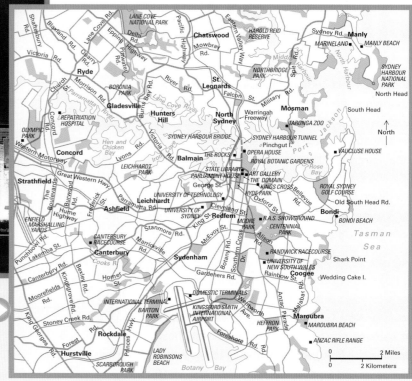

Sydney and its suburbs cover about 4,690 square miles (12,100 square kilometers). Downtown Sydney stands on the south side of Sydney Harbour. East of the downtown area lies a series of parks, including the Royal Botanic Gardens. The parliament house for New South Wales is also in this area. Suburbs spread out in every direction from the city.

Melbourne

Melbourne lies on Port Phillip Bay on Australia's southeastern coast. John Batman, an Australian farmer, founded Melbourne in 1835. He came from the nearby island of Tasmania, seeking land for sheep farms. He bought 600,000 acres (243,000 hectares) from the Aborigines and paid them with blankets, tomahawks, and other goods. Today, the city and its suburbs cover about 3,400 square miles (8,806 square kilometers).

Only a small portion of Melbourne's people are descended from the Aborigines. About one-third of Melbourne's population was born outside Australia. Many of these immigrants came from the United Kingdom and Ireland. Other immigrants include those from Italy and Greece. Australia's Asian community represents both East and Southeast Asia. Each ethnic group has introduced its own type of food, entertainment, and clothing, making Melbourne a truly cosmopolitan city.

Melbourne accounts for about 33 percent of all of Australia's factory output. The city's chief manufactured products include automotive equipment, clothing, coal, machinery, metal, paper, petroleum products, shoes, and textiles. Wholesale and retail trading and important financial institutions make Melbourne a major commercial center.

Memorial Gardens lies in the heart of Melbourne. A tree-lined pathway leads into the city center. Residents and visitors can enjoy numerous parks located throughout the city.

HISTORY

Australia's first settlers, the ancestors of today's Aborigines, probably reached the continent at least 50,000 years ago. The Europeans discovered Australia much later. They first entered the area during the 1500's, when Portuguese and Spanish explorers landed in New Guinea. These explorers and their successors were searching for a mysterious land they believed lay south of Asia. They called the continent Terra Australis Incognita, Latin for *Unknown Southern Land*.

In 1606, a Dutch navigator named Willem Jansz briefly visited what he thought was the coast of New Guinea. Actually, it was the coast of extreme northeastern Australia. Jansz thus became the first European known to sight the continent and land in Australia. Between 1616 and 1636, other Dutch navigators explored Australia's west, southwest, and north coasts.

In 1642, Abel Janszoon Tasman, a Dutch sea captain, sailed around the continent and briefly visited a land mass that he thought was part of the continent. Actually, it was an island. He named it Van Diemen's Land, but it was later named Tasmania in his honor. Finally, in 1770, James Cook of the British Royal Navy became the first European to see and explore Australia's fertile east coast. He claimed the region for Great Britain (later the United Kingdom) and named it New South Wales.

The Australian bicentennial, or 200th anniversary, was celebrated in 1988. In 1788, British ships reached New South Wales, and the passengers established the first British settlement in Australia.

Settlement and exploration

Before the American Revolution (1775–1783), Great Britain shipped many convicts to the American Colonies to relieve overcrowding in British jails. After the American Colonies won their independence, Britain had to find a new place to send convicts. In 1786, the British decided to start a prison colony in New South Wales. Captain Arthur Phillip was appointed to establish the colony and serve as its governor. He sailed from England in May 1787 with 11 ships carrying about 750 convicts, about 200 British soldiers, about 30 soldiers' wives, and a few children.

Phillip's ships reached Botany Bay, on Australia's east coast, in January 1788. He settled his group near a large harbor north of Botany Bay and thus founded the first European settlement in Australia. This was the beginning of the city of Sydney.

When New South Wales was first settled, no one knew whether Australia consisted of one huge land mass or of two or more large islands. From 1801 to 1803, Matthew Flinders sailed around the continent. He proved the mainland to be one vast land mass.

In 1829, a British sea captain named Charles Fremantle landed on Australia's southwest coast and claimed the entire western part of the continent for the United Kingdom. Later that year, Sir James Stirling officially established the colony of Western Australia and founded its capital, Perth. In 1860 and 1861, Robert O'Hara Burke and William Wills became the first Europeans to cross the continent from south to north.

TIMELINE

c. 45,000–60,000 B.C.	Ancestors of Aborigines settle in Australia.
1606	Willem Jansz becomes first known European to land in Australia.
1616-36	Dutch navigators map west, southwest, and north coasts.
1642	Abel Tasman discovers Van Diemen's Land (renamed Tasmania in 1855).
1770	James Cook explores Australia's east coast and claims it for Great Britain as New South Wales.
1788	Great Britain establishes prison colony in New South Wales.
1801-03	Matthew Flinders sails around Australia, proving it to be one land mass.
1829	Charles Fremantle claims western Australia for the United Kingdom.
1830-40's	First long expeditions into the interior begin.
1851	Colony of Victoria established. Gold discovered in New South Wales and Victoria.
1859	Colony of Queensland established.
1860-61	Burke and Wills cross the continent from south to north.
1868	The United Kingdom ends transportation of convicts to Australia.
1901	Australia becomes the Commonwealth of Australia, an independent nation.
1914-18	Australians join the United Kingdom in World War I (ANZAC forces).
1939-45	Australians fight in World War II.
1945	Australia becomes member of United Nations.
1965-73	Australian troops serve with U.S. troops in Vietnam War.
1967	Australian Constitution amended to permit the establishment of federal programs to aid the Aborigines.
1978	The Northern Territory becomes responsible for its own administration, the first step toward statehood.
1988	Australia celebrates its bicentennial.
2000	Sydney hosts the Summer Olympic Games.
2009	The worst wildfires in Australian history devastate southern Victoria, leaving 173 people dead.
2010-11	Flooding caused by heavy rains covers three-quarters of Queensland and one-quarter of Victoria in one of Australia's worst natural disasters.

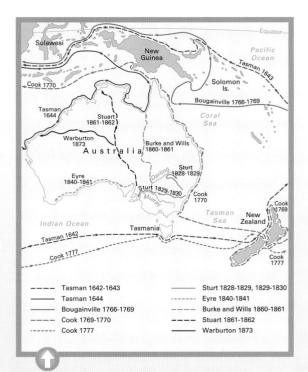

Tasman 1642-1643
Tasman 1644
Bougainville 1766-1769
Cook 1769-1770
Cook 1777
Sturt 1828-1829, 1829-1830
Eyre 1840-1841
Burke and Wills 1860-1861
Stuart 1861-1862
Warburton 1873

Australia had its coastline first explored by the Dutch. In 1770, James Cook claimed Australia's east coast for Great Britain. Later, bold adventurers tried to cross the continent's interior.

Sir Donald Bradman
(1908-2001)
Cricket player

Patrick White
(1912-1990)
Nobel prize author

Hugh Jackman
(1968-)
Actor

Becoming a nation

After the explorers and convicts came the sheep farmers and the prospectors. The farmers established rich grazing lands and became some of the largest landholders in Australia. With the discovery of Australian gold in the 1850's, the continent's population grew from about 400,000 to more than 1.1 million. In 1868, the United Kingdom ended its practice of sending convicts to Australia. As the number of free settlers in Australia grew, so did the colonists' demands for self-government. By 1890, all the Australian colonies had been granted self-government.

However, a growing number of Australians believed that the colonies would be better off as a single nation with a unified government. In 1897 and 1898, a federal convention drew up a constitution for Australia. The people approved it in balloting during 1898 and 1899, and the United Kingdom approved it in 1900. On Jan. 1, 1901, the six colonies became the states of a new nation—the Commonwealth of Australia.

REGIONS AND STATES

Australia has six states, two mainland territories, and eight territories outside the mainland. Canberra, the national capital of Australia, is in the Australian Capital Territory (ACT) in southeast Australia. Planners estimate that only about a third of the ACT is ever likely to be developed, but Canberra is the nation's leading example of large-scale city planning. The economy is dependent on the activities of the national government, which employs about half of the city's workers.

The Northern Territory occupies almost a sixth of the Australian continent. Although it was granted self-government in 1978, the Northern Territory has not yet been granted state status. The territory is known for its magnificent landscape and its mineral wealth. Nearly half the territory's people live in Darwin, and almost one-third of its people are Aborigines.

Queensland, in the northeastern corner of the Australian continent, is the second largest state in Australia. Nearly half of Queensland's people live in the Brisbane area, and most of the remainder live in the larger cities and towns along the eastern coast.

Southeastern Queensland produces large quantities of wheat and other grains, and Queensland leads Australia in the production of beef cattle. Queensland also has productive copper, lead, silver, and zinc mines.

New South Wales, south of Queensland, is the oldest state in Australia and has the largest population. It is also the richest and one of the most developed Australian states.

Most people in New South Wales live in the three main cities—Sydney, Newcastle, and Wollongong. Service industries employ about 75 percent of the state's labor force. Factories manufacture clothing and textiles, fertilizers, electronic products, steel, and copper.

Queensland lies within the tropics and subtropics. Some plateau and mountain areas receive a high rainfall and support lush forests.

Tasmania has a variety of scenery and historic relics that attract many tourists.

Kata Tjuta, also called the Olgas, is a group of more than 30 gigantic, dome-shaped rocks west of Uluru in the Northern Territory. Many people consider these colorful rocks to be even more spectacular than Uluru.

Perth is the capital and chief business center of the state of Western Australia. The downtown area stands on the northern bank of the Swan River. Modern high-rise buildings dominate the skyline.

Victoria, in the southeastern corner of the Australian continent, is the smallest and most densely populated of the mainland states. Victoria is a highly industrialized, highly urbanized area, and most of its people work in manufacturing and commerce. Melbourne, Victoria's capital, is one of the 100 largest cities in the world. Most of the state's population and industry are concentrated there. Rich crop- and dairy-farming lands lie in the southeast, while the southwest has extensive grazing lands. In the north, these two areas merge into wheat-farming plains.

Tasmania, the island state of Australia, is the smallest state and also one of the most beautiful. Many Australians enjoy vacations in Tasmania. Lakes and rivers in the mountains of the island's interior supply electricity for industry. The coastline has many scenic capes and bays. Tasmania's mountainous west coast is rich in copper, tin, and iron ore.

South Australia is the third largest state in Australia. It occupies a central position on the southern coast and covers one-eighth of the country. Adelaide, where three-fourths of the state's people live, is an important industrial and cultural center. Although South Australia is very dry and must be irrigated, it produces barley, wheat, and wine. Its iron-ore deposits are a major source of raw materials for the nation's iron and steel industries. South Australia is the country's main source of salt. Manufacturing industries account for about half of South Australia's economy.

Western Australia, the largest of the six Australian states, occupies the western third of the country but has only about 10 percent of the people. Large areas of desert and semi-desert cover its inland regions. About 75 percent of the people of Western Australia live in the Perth region, the only large urban area in Western Australia. Iron ore is the state's most important mineral, but Western Australia is known especially for its rich deposits of gold and diamonds.

Australia's outlying territories are Ashmore and Cartier Islands, Australian Antarctic Territory, Christmas Island, Cocos Islands, Coral Sea Islands, Heard and McDonald Islands, Jervis Bay, and Norfolk Island.

PEOPLE

The great majority of the Australian people belong to the middle class. Most Australians also have similar educational backgrounds and hold similar values and attitudes. The ways of life throughout the country are therefore remarkably uniform. Even the differences between life in the cities and life in rural areas are relatively minor.

Ancestry

Most Australians are European immigrants or descendants of European immigrants. Aborigines make up only about 2 percent of the population. Traditionally, Australia has relied heavily on immigrants to build up its labor force. Through the years, millions of immigrants have been attracted to Australia by the promise of high-paying jobs.

Australia has admitted about 6-1/2 million immigrants since the 1950's. Many of the newcomers have come from the United Kingdom and Ireland. Most of the others came from mainland Europe, especially the former Yugoslav republics and such southern European countries as Greece and Italy. Since the 1970's, however, the number of immigrants from New Zealand and Southeast Asia has increased rapidly.

Way of life

About 9 out of 10 Australians live in cities and towns, making Australia one of the world's largest urbanized countries. About 75 percent of all Australians live in cities of more than 100,000 people. Most city dwellers live in suburbs, the residential areas that extend outward from the central business district. Most families live in single-story houses, each with its own garden. Most families own their own home. Australian cities have few apartment buildings.

English is Australia's official language. Australian English includes many British terms but differs from British English in certain ways. British settlers had to develop a vocabulary to describe the many unfamiliar animals and plants

The Aborigines were the first Australians. Today, some Aborigines live in tribal settlements and preserve traditional ways of life. Many Aborigines lag far behind other Australians in both education and income.

Cricket, an English game played with a bat and ball, is a favorite summer sport in Australia. The Australian national cricket team regularly plays against teams from other countries.

Surf lifesavers stand ready to demonstrate the use of a surf reel as a rescue method at one of Australia's popular surf carnivals. Australia was the first country in the world to develop a surf lifesaving movement, which consists of trained voluntary lifesavers who patrol ocean beaches on weekends and holidays to make them safe for swimmers. Surf carnivals are colorful and spectacular competitions held by surf lifesaving clubs to promote the public's interest in the lifesaving movement.

A variety of cuisines is available in Australia. Greek, Italian, and various other European styles of cooking have become increasingly popular as the number of immigrants from those countries has increased. Many Australians have also developed a taste for Chinese, Indonesian, and Vietnamese foods.

in their new environment, and, in some cases, they borrowed words from the Aborigines. For example, kangaroo and koala are Aborigine words. Pioneer settlers in the Australian interior also invented a large, colorful vocabulary. Ranches became known as stations, wild horses as brumbies, and bucking broncos as buckjumpers. Understanding Australia's most famous song, "Waltzing Matilda," requires translation of many local terms: a matilda is a blanket roll, to waltz matilda means "to tramp the roads," a swagman is a tramp, and so on.

Each Australian state and the Northern Territory has its own laws concerning education. The federal government regulates education in the rest of the country—that is, in the Capital Territory and other territories.

Australian children attend primary schools for six to eight years, depending on the state or territory. Australian secondary schools offer five or six years of education. However, about one-fourth of the students leave school by the time they reach the age requirement, and they may complete only three or four years of secondary education. Most students who graduate from secondary school go on to a university or college.

The Australian Constitution forbids a state religion and guarantees religious freedom. The majority of Australians are Christians, but many do not attend church regularly. Roman Catholics make up about one-fourth of the population.

Outdoor sports are extremely popular in Australia. Many people enjoy skin diving, surfing, swimming, and boating, as well as golf and tennis. Team sports are a national pastime. Australians begin to play team sports in elementary school, and many continue to enjoy them throughout life. The best players may work their way up through local and state competitions and even win a position on one of the national teams. Australia's professional sports teams have large and enthusiastic followings.

THE GREAT BARRIER REEF

Along the northeast coast of Australia lies the Great Barrier Reef, the largest group of coral reefs in the world. This chain consists of more than 3,000 reefs and many small islands, and it extends for about 1,400 miles (2,300 kilometers). In some sections, it lies more than 100 miles (160 kilometers) from the Australian coast. Other parts lie only about 10 miles (16 kilometers) out.

A special ecosystem

A coral reef is a limestone formation that lies under the sea or just above the surface. The coral that forms the Great Barrier Reef is made up of hardened skeletons of dead water animals called polyps. Billions of living coral polyps are attached to the reef. The polyps are extremely colorful, as are the many sea animals that live in the Great Barrier Reef. Together, they create a beautiful sea garden.

The Great Barrier Reef supports about 400 species of polyps, hundreds of species of fish, and more than 200 kinds of birds. Crabs, giant clams, and sea turtles also live on the reef. The warm waters around the reefs and the beauty of the coral formations attract swimmers, skin divers, and tourists from all around the world.

Protecting the reef

The reef region has long been used as a source of food and raw materials. For example, such sea animals as scallops, shrimp, and fish are abundant there. The Great Barrier Reef, or parts of it, were explored and used by Aboriginal fishermen and hunters many thousands of years ago. Since the arrival of Europeans in 1788, the reef has supported business ventures. The prospect of finding oil in the region has attracted petroleum companies that want to drill in the area. During the late 1960's, however, public concern grew about damage to the reef from these and other sources.

A scuba diver is attracted to the warm waters and beautiful fan coral formations of the Great Barrier Reef. These corals have strong, flexible skeletons that branch to form a lacy network.

An aerial view of the Great Barrier Reef shows the extreme clarity of the warm tropical water. Visitors can get a close-up view of the reefs with scuba and snorkeling equipment, glass-bottomed boats, and underwater observatories.

In 1970, two government commissions were established to investigate the risks of oil drilling in Great Barrier Reef waters. Following the investigation, the Australian government made most of the reef a national park. The Great Barrier Reef Marine Park Authority (GBRMPA), an Australian government agency, works to protect the reef from damage. The government has made it illegal to collect any of the coral. Mining and oil drilling are prohibited in the Great Barrier Reef region.

In 2004, the GBRMPA began to implement a zoning plan that created a network of wildlife sanctuaries. The plan prohibited all commercial activities except for tourism in one-third of the reef's area.

Researchers continue to monitor factors that threaten the health of the Great Barrier Reef. Researchers look at the impact human diving activity has on the area and how that impact can be minimized. In addition, researchers examine the effects of global warming, of pollutants that may seep into the area, and of naturally occurring threats such as diseases and the crown-of-thorns starfish. These starfish feed on living polyps and can destroy up to 95 percent of the corals on any single reef.

The colorful sea slug belongs to the nudibranch (sea snails) family. Frilly outgrowths from the body, called cerata, are thought to be respiratory organs.

Hundreds of species of fish find shelter in the hard coral of the Great Barrier Reef. Coral reefs sustain more species of fish than any other marine environment. Other animals that live on the reefs include crabs, giant clams, and sea turtles.

AUSTRIA

Austria is a small, landlocked country in central Europe. It shares boundaries with Switzerland and Liechtenstein to the west; Germany, the Czech Republic, and Slovakia to the north; Hungary to the east; and Slovenia and Italy to the south.

Austria's scenic beauty attracts millions of visitors each year. Visitors to the Austrian Alps enjoy skiing and a host of other winter sports. In addition to the majestic snow-capped Alps and their foothills—which stretch across the western, southern, and central parts of the country—Austria has many crystal-clear lakes, and thick forests cover much of the land. Many of Austria's picturesque villages nestle in the broad, green valleys.

People also come to Austria to enjoy the many impressive sights and sounds of one of the great cultural centers of Europe. Austria has made outstanding contributions in the fields of architecture, literature, painting, and, above all, music. Joseph Haydn, Wolfgang Amadeus Mozart, Johann Strauss, and Johann Strauss, Jr., are among the great composers Austria has produced, and the country's musical tradition continues today, as thousands flock to its concerts, operas, and music festivals.

Vienna, Austria's capital and largest city, is also the country's leading cultural, economic, and political center. About a fifth of Austria's people live in Vienna. The city's historical section contains many art galleries, churches, theaters, and other beautiful buildings. Many people gather at the sidewalk cafes to enjoy the view and sample some of the delicious pastries and coffee with whipped cream for which Vienna is famous.

Although Vienna lost much of its political importance after World War I (1914-1918), its many landmarks, including the Schönbrunn Palace, recall Austria's former position as one of the most powerful countries in Europe. Austria was the center of a huge empire ruled by the royal Habsburg family from the 1200's until the empire collapsed in 1918. Austria then became a republic, and a long period of economic difficulty and political unrest followed.

By the early 1950's, however, industry had increased, and Austria's economy began to recover. The country also achieved political stability. Today, as a neutral nation, Austria often serves as a meeting place where representatives of different countries gather to exchange ideas. For example, Vienna hosted meetings of the Strategic Arms Limitations Talks (SALT) between the United States and what is now the former Soviet Union. In addition, the Austrian capital is home to several United Nations agencies. In a nationwide election held in June 1994, Austrians voted to join the European Union (EU). The country became an EU member on Jan. 1, 1995.

AUSTRIA TODAY

Austria is a federal republic. Its national government, which was formed after World War II (1939-1945), is based on the democratic Constitution adopted in 1920.

The president, Austria's head of state, is elected by the people to a six-year term. The president's duties are largely ceremonial and include appointing ambassadors and acting as commander in chief of the armed forces.

The chancellor, or prime minister, and Cabinet run the Austrian government. The president appoints the chancellor, who is usually the leader of the political party with the most seats in the *Nationalrat* (National Council).

The chancellor selects the Cabinet members to head government departments. Government policies formed by the chancellor and the Cabinet must be approved by Parliament.

Austria's Parliament is made up of two houses: the Nationalrat, the lower house, and the *Bundesrat* (Federal Council), the upper house. The 183 members of the Nationalrat are elected by the people, and the 62 members of the Bundesrat are elected by the country's nine *Landtags* (provincial legislatures).

The land

Austria's six main land regions are the Granite Plateau, the Eastern Forelands, the Alpine Forelands, the Northern Limestone Alps, the Central Alps, and the Southern Limestone Alps. Austria's alpine regions have a beautiful scenic landscape of rolling hills, forested mountain slopes, and rugged peaks, dotted with large glaciers and sparkling lakes. The Granite Plateau in northern Austria is a region of hills and mountains that consist mostly of granite. Dense forests cover part of this area.

Most of Austria's land is too mountainous for raising crops. The country's chief agricultural area—a lowland called the Vienna Basin—is located in the northern part of the Eastern Forelands. Vienna, Austria's capital and largest city, stands in this region.

FACTS

Official name:	Republik Osterreich (Republic of Austria)
Capital:	Vienna
Terrain:	In the west and south, mostly mountains (Alps); along the eastern and northern margins, mostly flat or gently sloping
Area:	32,383 mi² (83,871 km²)
Climate:	Temperate; continental, cloudy; cold winters with frequent rain in lowlands and snow in mountains; cool summers with occasional showers
Main rivers:	Danube, Drava, Inn, Enns, Mur
Highest elevation:	Grossglockner, 12,461 ft (3,798 m)
Lowest elevation:	Neusiedler Lake, 377 ft (115 m) below sea level
Form of government:	Federal republic
Head of state:	President
Head of government:	Chancellor
Administrative areas:	9 bundeslaender (states)
Legislature:	Bundesversammlung (Federal Assembly) consisting of Bundesrat (Federal Council) with 62 members serving terms of varying lengths and the Nationalrat (National Council) with 183 members serving four-year terms
Court system:	Oberster Gerichtshof (Supreme Judicial Court); Verwaltungsgerichtshof (Administrative Court); Verfassungsgerichtshof (Constitutional Court)
Armed forces:	34,900 troops
National holiday:	National Day - October 26 (1955)
Estimated 2010 population:	8,406,000
Population density:	260 persons per mi² (100 per km²)
Population distribution:	67% urban, 33% rural
Life expectancy in years:	Male, 77; female, 83
Doctors per 1,000 people:	3.7
Birth rate per 1,000:	9
Death rate per 1,000:	9
Infant mortality:	4 deaths per 1,000 live births
Age structure:	0-14: 15%; 15-64: 68%; 65 and over: 17%
Internet users per 100 people:	59
Internet code:	.at
Languages spoken:	German (official), Turkish, Serbian, Croatian, Slovene, Hungarian
Religions:	Roman Catholic 73.6%, Protestant 4.7%, Muslim 4.2%, other 17.5%
Currency:	Euro
Gross domestic product (GDP) in 2008:	$421.37 billion U.S.
Real annual growth rate (2008):	1.6%
GDP per capita (2008):	$51,255 U.S.
Goods exported:	Iron, machinery, paper products, steel, vehicles
Goods imported:	Chemicals, food, machinery, petroleum, vehicles
Trading partners:	Czech Republic, Germany, Italy, Switzerland, United States

The people

Most Austrians live in the Eastern Forelands and in the area just south of the Danube River. About 67 percent of the people live in urban areas, such as the major cities of Vienna, Graz, Innsbruck, Linz, and Salzburg.

The people of Austria enjoy good food. A Viennese dish called *Wiener schnitzel* (breaded veal cutlet) has become a favorite in many countries, and the delicious cakes and pastries created by Austrian bakers are world renowned. Austrians also love the outdoors. They enjoy a variety of outdoor sports throughout the year in the country's many forests, lakes, and mountains.

The Republic of Austria is divided into nine provinces: Burgenland, Carinthia, Lower Austria, Salzburg, Styria, Tyrol, Upper Austria, Vienna, and Vorarlberg. About three-fourths of the country is covered by mountains, and much of the remainder consists of rolling hills and broad valleys. Most of Austria's people live in the eastern part of the country just south of the Danube River.

Austria's Parliament Building in Vienna, was completed in 1883. The building was designed in the classical Beaux Arts style, with rows of impressive stone columns. A 13-foot (4-meter) statue of Athena, the Greek goddess of warfare and wisdom, stands in front of the Parliament Building. The clock tower atop Vienna's City Hall, or Rathaus, rises in the background.

HISTORY

Celtic tribes moved into central and eastern Austria around 400 B.C. By 15 B.C., the Romans controlled the country south of the Danube. After the West Roman Empire collapsed in A.D. 476, many different peoples invaded Austria.

Finally, in 955, Austria came under the rule of Otto I, the king of Germany. In 962, the pope crowned Otto emperor of what later became known as the Holy Roman Empire. German emperors ruled the Holy Roman Empire until it ended in 1806.

The Habsburg Empire

The Babenberg family controlled northeastern Austria from 976 until 1246, when the last Babenberg duke died without an heir. The king of Bohemia then seized the region. In 1273, a member of the Swiss Habsburg family became Holy Roman Emperor Rudolf I. Rudolf claimed the Babenberg lands and defeated the Bohemian king at the Battle of Marchfeld in 1278.

The Habsburgs lost the Holy Roman crown in the 1300's, but a Habsburg was once again elected emperor in 1438. From then on, the Habsburgs held the title almost continuously.

In the 1400's and 1500's, the Habsburg emperors acquired new lands, including Bohemia and Hungary. However, their authority was shaken by the Protestant Reformation in the 1500's and the Thirty Years' War of 1618 to 1648. During these years, Austria emerged as the chief state in the empire. The Ottoman Turks tried to drive Austria from Hungary and made two attacks on Vienna, but they were defeated in the late 1600's.

In 1740, Empress Maria Theresa also fought to maintain her possessions. When the fighting ended in 1748, Maria Theresa lost Silesia to Prussia.

In 1806, after suffering several defeats in the Napoleonic Wars of the late 1700's and early 1800's, Emperor Francis II was forced to dissolve the Holy Roman Empire.

During the 1800's, revolutions broke out across Europe. Austria's minister of foreign affairs, Prince Klemens von Metternich, tried to suppress all revolutionary movements in the Austrian Empire, but in 1848, revolutionaries demanded the establishment of a constitutional government. Metternich fled.

Although the Austrian army had put down all revolts by 1851, the empire weakened in the years that followed. Austria lost its land in Italy to Italian and French forces, and Prussia replaced Austria as the leader of the German states in Europe. In 1867, Austrian Emperor Franz Joseph was forced to give equal status to his Hungarian holdings and create the Dual Monarchy of Austria-Hungary.

World wars

In the late 1800's and early 1900's, Slavs in Austria-Hungary demanded the right to govern themselves. Then, in 1914, Gavrilo Princip, a member of the Slavic nationalist movement in Serbia, killed Archduke Franz Ferdinand, the heir to the Austro-Hungarian throne. In response, Austria-Hungary declared war on Serbia, thus starting World War I. Germany and other countries joined Austria-Hungary in fighting the Allies—the United Kingdom, France, Russia, and, eventually, the United States.

In 1918, Austria-Hungary was defeated. The last Habsburg emperor was overthrown, and the empire was split into several countries. Austria became a republic. It adopted a democratic constitution in 1920, but conflicting political parties struggled for supremacy. In 1934, members of the Austrian Nazi Party killed Chancellor Engelbert Dollfuss, and in 1938, German troops seized Austria. Adolf Hitler united Austria and Ger-

Empress Maria Theresa of Austria was one of Europe's most powerful rulers during the second half of the 1700's.

The defeat of the Turks at Vienna in 1683 is depicted in the painting. An army led by Austria's Duke of Lorraine conquered a Turkish army of more than 200,000 men. Austria's success was seen as a victory of Christianity over Islam.

TIMELINE

400 B.C.	Celtic tribes occupy Austria.
15 B.C.	The Romans control Austria south of the Danube River.
A.D. 100's	Invasions by tribes from the north weaken Roman control.
976	The Babenberg family controls northeastern Austria.
1156	Vienna becomes capital.
1278	Rudolf I, a Habsburg, begins to acquire the Babenberg lands for his family.
1438-1806	The Habsburgs hold the title of Holy Roman emperor almost continuously.
1453	The Duchy of Austria becomes the Archduchy of Austria.
1683	Austria defeats the Turks at Vienna.
1740	Maria Theresa inherits her father's lands and the War of the Austrian Succession begins.
1740-1748	Maria Theresa loses Silesia to Prussia in the War of the Austrian Succession.
1806	Holy Roman Empire is dissolved.
1814-1815	At the end of the Napoleonic Wars, the Congress of Vienna returns to Austria most of the land it had lost during the wars.
1848	Revolutionaries in Vienna demand the establishment of a constitutional government.
1867	Austria-Hungary is established.
1914-1918	Austria-Hungary is defeated in World War I.
1920	Austria adopts a democratic constitution.
1938	Adolf Hitler unites Austria with Germany.
1939-1945	The Allies defeat Germany in World War II.
1945-1955	The Allies occupy Austria.
1955	Austria agrees to be permanently neutral in international military affairs.
1987	The Socialist Party forms a new coalition with the People's Party.
1995	Austria joins the European Union.

Prince Klemens von Metternich (1773-1859) Statesman

Wolfgang Amadeus Mozart (1756-1791) Composer

Franz Joseph (1830-1916) Emperor, 1848-1916

The Habsburg empire grew from lands acquired in the late 1200's by Rudolf I. By 1526, the Habsburg family had taken control of large parts of Bohemia and Hungary. Over the next four centuries, the borders and dominant areas within the empire shifted. At the end of World War I, the empire ended and the republic of Austria was born.

many and led both countries into World War II in 1939.

After Germany was defeated in 1945, Austria was occupied by the Allies, and a government based on Austria's 1920 Constitution was established. In 1955, the Allies ended their occupation with the understanding that Austria would remain permanently neutral in international military affairs.

Since the 1950's, Austria's economy has grown steadily, and the country has been politically stable. As a neutral nation, Austria has been the site of many international diplomatic meetings.

SOUTH AND WEST PROVINCES

Several high ranges of the Alps, separated by beautiful river valleys, cross southern and western Austria. The southern and western provinces consist of Vorarlberg, Tyrol, Salzburg, and Carinthia. Innsbruck and Salzburg are the chief cities.

Vorarlberg

Vorarlberg is a small province in the western corner of Austria. Vorarlberg's ancient capital, Bregenz, is a charming town on the eastern shore of Lake Constance. Summer festivals held on the lake attract many visitors each year.

Much of the land in Vorarlberg is too mountainous for raising crops, but dairy animals graze in the high areas and yield large quantities of milk and cheese. Skilled craftworkers in the region produce fine embroidery, and other goods include watches, clocks, metals, chemicals, and pharmaceuticals. Tourism is also a leading industry in the province, and sports centers draw visitors to the province the year around.

The people of Vorarlberg, who speak a German dialect known as *Alamannic,* have formed close relationships with the Alamannic-speaking populations of the Allgäu region of Bavaria in southwestern Germany, the Lake Constance area of eastern Switzerland, and Liechtenstein. As a result, the province has established close economic ties with these regions, and many people from Vorarlberg commute across international borders every day to work in Liechtenstein or in St. Gallen, Switzerland.

Tyrol

After World War I, the region of Tyrol was divided into two parts: Northern Tyrol, which was given to Austria, and Southern Tyrol, which was given to Italy. Austrian Tyrol, or Northern Tyrol, lies in the mountainous western part of the country.

One of the most popular holiday areas in Europe, Tyrol offers superb facilities for summer and winter sports. The Alps cover most of the province, and Grossglockner, the highest mountain in Austria, rises 12,461 feet (3,798 meters) in this region. The wide, fertile Inn River Valley extends over the northern part of Tyrol.

Innsbruck, the capital city of the province, lies in the Alps north of Brenner Pass. This Alpine pass links northern Europe with the Mediterranean countries. Innsbruck's many beautiful buildings include the Hofkirche, a church that contains the tomb of Maximilian I, the Holy Roman emperor from 1493 to 1519.

The province is Austria's favorite holiday destination, and tourism is Tyrol's main economic activity. Other industries in the region produce leather, processed foods, stained glass, and textiles.

Salzburg and Carinthia

The province of Salzburg lies in the valley of the Salzach River. The region produces salt (for which it is named), as well as leather, paper, textiles, and timber. Agricultural activities in the area include dairy farming and horse breeding.

Salzburg is also the cultural heart of central Austria. Its capital city, also named Salzburg, is the birthplace of Wolfgang Amadeus Mozart, one of the world's greatest composers. Annual music and theater festivals draw people from all over the world to this beautiful city on the Salzach River. Salzburg's many magnificent buildings include Hohensalzburg Castle, set high on a hill overlooking the historic section of the city, and the Residenz, once the palace of Salzburg's prince bishops. Other important buildings include Salzburg's baroque cathedral, built between 1614 and 1628, and Mozart's birthplace.

Carinthia lies in south-central Austria in a sheltered basin surrounded by mountains. The region is known as the country's "sun terrace" because of its comparatively warm climate. Forests cover more than half of the province.

Carinthia's warm summers attract numerous visitors to its resort areas, including beautiful Lake Wörther. Another popular attraction is the splendid Schloss Hochosterwitz, a castle perched on a cliff near St. Veit.

Tourism plays a major part in the local economy, but agriculture, dairy farming, forestry, mining, and paper production are also important industries. The small town of Ferlach is famous for its manufacture of high-quality hunting rifles.

Innsbruck the old provincial capital of Tyrol, retains many of its medieval buildings, narrow streets, and tall houses. It is home to a university and several industries.

Hohensalzburg Castle overlooks the old section of the city of Salzburg. The castle, which is more than 900 years old, was rebuilt in the 1500's.

The church of St. Maria Wörth lies on the Wörther See, the largest of the Alpine lakes of Carinthia.

NORTH AND EAST PROVINCES

The Danube River flows across a varied area of hills, mountains, and valleys in northeastern Austria. The north and east provinces include Upper Austria, Lower Austria, Vienna, Burgenland, and Styria. The chief cities are Vienna, Graz, and Linz.

Upper Austria

Upper Austria, a province in northern Austria, is spanned by the Danube River. Linz, its capital, is Austria's third largest city, and it is the country's most important heavy industrial center. Manufactured items include ball bearings, engineering products, steel, and vehicles.

The province also contains many natural resources. Upper Austria's farmland yields cereals, fruit, potatoes, and sugar beets, while the Alpine foothills support livestock production. Large forests cover the higher elevations. The ancient salt mines in the lake district of Salzkammergut are still productive, and the Danube and Enns rivers generate power for large hydroelectric stations.

Upper Austria's natural beauty also contributes to the economy. Tourists flock to the province's Alpine region, and many people visit the area's spas and health resorts.

Lower Austria and Vienna

Lower Austria, the country's largest province, lies in the northeastern corner of Austria. It completely surrounds the small province of Vienna. Vienna is Austria's capital and largest city, as well as its smallest province in area. The two provinces were separated in 1922, but Vienna is still the focal point of both. It is Austria's chief industrial city and a leading European cultural center.

The main towns in Lower Austria include St. Pölten (its capital), Wiener Neustadt, and Baden. The province has one of the most varied landscapes in Austria, ranging from high mountains and wooded hills to fertile valleys and grasslands. Due to efficient agricultural methods, the available farmland yields sugar beets, wheat, and wine grapes. Farmers also raise livestock.

Iron miners have worked in the Erzberg, near Eisenerz in Styria, since the Middle Ages. The Erzberg has one of the largest iron ore deposits in Europe.

The Wachau Valley, stretching along the Danube between Melk and Krems in Lower Austria, is widely considered to be the most beautiful part of the Danube. The Bohemian Forest lies to the northwest, and the Dunkelsteiner Forest to the southeast. The region contains quaint old towns surrounded by vineyards and overlooked by historic castles.

Wilhering Abbey has an ornate interior with superb frescoes that date from the 1700's. The abbey, which stands near Linz on the banks of the Danube, was founded in 1146.

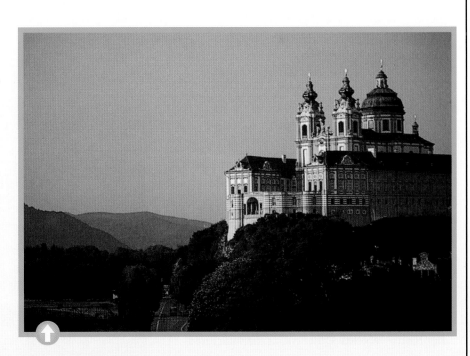

The Benedictine Abbey of Melk stands high above the banks of the Danube River in Lower Austria. Built between 1702 and 1738, the abbey occupies the former site of an ancient castle.

Heavy industry in the region south of the fertile Vienna Basin includes chemical plants, iron and steel works, and textile- and food-processing industries. A number of large hydroelectric power stations lie along the Danube, which runs through the middle of the province. Tourism is also an important industry in Lower Austria, and resorts in the province's eastern Alpine region attract many visitors.

Burgenland

Burgenland, the most easterly of Austria's provinces, is composed of lands that shifted between Austria to Hungary from the 1400's to 1647, when Hungary took control. Austria acquired the territory from Hungary in 1921, when four regions were united to form Burgenland. However, one of the regions was soon returned to Hungary.

The province's landscape includes Alpine foothills as well as the fringes of the Upper Hungarian Lowlands, a popular vacation area. The picturesque castles and fortresses that line Neusiedler Lake, the lowest point in Austria, also attract many visitors.

Much of the province lies in the Vienna Basin, whose fertile soil helps make Burgenland the country's chief agricultural area. Farmers in the area grow corn, fruit, grapes, sugar beets, and vegetables. Even the industries in the province—sugar refining and canning—reflect its agricultural importance.

Styria

Styria, in the southeast, is Austria's second largest province. The area has a variety of landscapes and climates, ranging from the chilly Alpine regions of the north and central parts of the province to the warm, sunny plains of the south and west. Graz, which lies on the Mur River, is the capital of Styria and Austria's second largest city. The city contains many historic buildings, including the Charles Francis University, which dates from 1586, and a cathedral built in the 1400's.

Grapes, corn, and wheat are grown in Styria's fertile Alpine foothills, and dairy animals graze on the higher Alpine pastures. The timber industry flourishes in the thickly forested mountainous regions. Iron ore deposits mined from the Erzberg (Ore Mountain) near Eisenerz provide raw materials for the area's iron and steel works. Styria's rivers are harnessed as a source of hydroelectric power.

ECONOMY

Austria's economy was brought to a standstill as a result of World War II. After the war, with aid from the United States and other Western nations, the Austrian government bought up certain key industries, including coal and metal mining, electric power production, iron and steel production, and oil drilling and refining. Since the early 1950's, Austria has become increasingly industrialized, and its economy has grown steadily. Today, Austria is a prosperous country with little poverty or unemployment.

Austria's leading manufacturing activities are the production of metals, such as iron and steel, and metal products, including automobiles and other motor vehicles. Other major manufactured products include chemical products, electrical equipment, processed foods and beverages, and textiles and clothing. In addition, many smaller factories and workshops produce fine handicrafts, such as glassware, jewelry, needlework, porcelain objects, and woodcarvings.

Service industries

Service industries make an important contribution to the economy, accounting for the largest part of Austria's annual economic production. Community, government, and personal services employ about a fourth of the country's workers. Wholesale and retail trade form the most important service industry in terms of value of production. Other leading service industries include communications, finance, transportation, and utilities.

Tourism is also important to Austria's economy, and the nation is one of Europe's most popular vacation spots. Sports centers in the Alps attract many winter vacationers, particularly skiers, and the lake resorts in central Austria are popular in the summer. In addition, many people come to Austria to enjoy the museums and concert halls of Vienna, as well as the summer music festivals held throughout the country.

The Danube River, shown here flowing through Linz, is a major shipping route for trade between Austria and nearby countries. Passenger vessels also travel on the Danube. Most of Austria's large factories are located in the valleys of the Danube and other rivers.

Most farms in Austria are small, and all are privately owned. Some farmers—unable to earn enough from agricultural production—make extra money by opening their houses to tourists during the summer.

Wine grapes are cultivated in the warmer eastern provinces of Lower Austria, Burgenland, and Styria. More than half of Austria's grapes are grown along the Danube Valley northwest of Vienna on steep, terraced vineyards.

Skiers from many countries flock to Innsbruck, Kitzbühel, and other superb ski areas in the Austrian Alps. Austria is one of Europe's most popular vacation spots.

Agriculture and natural resources

By the 1970's, the development of manufacturing and service industries resulted in a sharp decrease in the number of people employed in agriculture. Nevertheless, Austrian farmers today supply more than 80 percent of the nation's food. Although only about 20 percent of the land is suitable for farming, modern machinery and farming methods have greatly increased production.

While the heart of Austria's cropland is the Vienna Basin, farms are found in every province. Dairy farming and livestock are the main sources of farm income, producing all the eggs, meat, and milk needed by the people. Austria's farmers also grow all of the country's potatoes and sugar beets and most of its barley, oats, rye, and wheat. Other farm crops grown in Austria include apples, barley, corn, grapes, hay, hops, pears, and vegetables. Farm animals graze in mountainous

areas, where the ground is too rugged and the climate too cold for crops. Wine making is also an important industry in Austria.

Although Austria is rich in mineral resources, many deposits are low-grade or too small to be profitably mined. The coal mines in Styria, for example, mainly yield a low-grade coal called *lignite*. As a result, Austria must import high-quality coal, as well as petroleum and natural gas.

Austria is one of the world's leading producers of magnesite. Other mineral deposits include gypsum, salt, stone, and tungsten. In addition, Austria's forests, which cover about 45 percent of the country, provide plentiful timber, paper, and other products. Swift-flowing rivers, perhaps the country's most important natural resource, provide energy for the hydroelectric power stations that produce most of the nation's electricity.

A MUSICAL NATION

Austria's strong musical tradition dates back more than 200 years. Many great composers, such as Joseph Haydn, Wolfgang Amadeus Mozart, and Franz Schubert were born in Austria, and during the 1700's and 1800's, Vienna in particular became an important center for the German-speaking music world. Ludwig van Beethoven, Johannes Brahms, Haydn, Mozart, Schubert, and Johann Strauss all lived in Vienna.

Plaques and memorials throughout the country testify to Austria's glorious musical past. In Vienna, where Haydn spent the last years of his life, visitors can see the Haydn Museum. Haydn developed the symphony and helped make it one of the most important forms of musical composition. Haydn and Mozart, who is also honored in Vienna, were the leading composers of the classical period of music, from the mid-1700's to the early 1800's. The rooms in Vienna where Mozart composed one of his operas, *The Marriage of Figaro* (1786), are open to the public, and Mozart's birthplace in Salzburg attracts music enthusiasts from all over the world.

Of course, Austria's composers are best remembered through their enchanting music, and the pleasure of relaxing at a Viennese sidewalk cafe is often enhanced by the accompanying strains of a Strauss waltz. The beautiful songs of Schubert, Gustav Mahler, and other great Austrian composers are regularly broadcast over Austrian radio.

The tradition continues

Austria's Arnold Schoenberg became one of the most revolutionary composers of the early 1900's. He developed a new system of composition called the *twelve-tone technique* and influenced many other composers, including fellow Austrians Alban Berg and Anton Webern.

Today, the country continues to make important musical contributions. The Vienna Boys' Choir, Vienna Philharmonic Orchestra, Vienna State Opera, and Vienna Symphony Orchestra have won international fame. In addition, Austria has many fine music schools, such as the Academy of Music in Vienna, which draws students from all over the world. The Vienna State Opera House presents operas to packed houses, and enthusiastic crowds throng Austria's many music festivals, which are held nearly all year-round.

Salzburg's Getreidegasse, the street where Mozart was born, attracts many visitors.

The simple table and chair at which Mozart worked are displayed at the Mozarteum, a music academy in Salzburg. Mozart composed his opera *The Magic Flute* in 1791, the year he died in poverty in Vienna.

Music festivals

Every year in May and June, Vienna hosts the Vienna Festival, a celebration of music, art, and theater. However, the emphasis is on music, and the city's concert halls are filled with the sounds of Austria's favorite composers. Opera is also highlighted at the festival in superb performances at the State Opera House and the Volksoper.

Festivals in Salzburg take place throughout the year. At the annual summer Salzburg Festival, a great musical and theatrical event, performances featuring works by Mozart are presented in the converted stables and riding school of the Archbishops of Salzburg. Another festival house was carved into the face of a mountain behind the stables. Other festivals include Mozart Week, which is held in January, and the Easter Music Festival, held during Easter week.

Music festivals also take place in other cities and towns throughout Austria. At the summer festival in Bregenz, concerts are performed on a floating stage on Lake Constance. A festival in Linz honors the works of Anton Bruckner, an Austrian composer of the middle and late 1800's who once lived in that city. Innsbruck stages the Tyrolean Summer Festival. Classical music offerings are presented in Innsbruck's concert halls, while brass bands, folk music, and dances can often be enjoyed outdoors.

The Golden Hall of the Vienna Concert Hall is the home of the Philharmonic Ball, a major annual event in the musical life of the city. The famous Vienna Philhartmonic Orchestra performs at the ball.

The Vienna State Opera House is internationally renowned for its impressive staging of operas and ballets.

A street musician plays the cello for donations from passersby in a passageway of the Hofburg, the old imperial palace in Vienna.

VIENNA

Vienna, Austria's capital city, lies in the northeastern part of the country on the south bank of the Danube River. The city's location at the intersection of a number of trade routes helped its growth. The Habsburgs made Vienna their capital in 1273, and the city's economic and political importance grew rapidly thereafter. Vienna was badly damaged during World War II, but the Viennese rebuilt almost all of the city's landmarks. Over time, the city regained much of its former spirit and wealth.

A modern city

Today, Vienna is Austria's chief industrial city. Its industries manufacture chemicals, clothing, leatherware, medicine, and radio and television products. The city also hosts several international agencies, including a United Nations (UN) center that serves as a conference site and provides office space for some UN agencies.

Vienna is one of the leading cultural centers in Europe. During the 1700's and 1800's, the city was a renowned center of literature, music, and science. A number of famous composers, scientists, and writers made their home in the city, among them Ludwig van Beethoven, Sigmund Freud, and Hugo von Hofmannsthal.

Today, tourists come from around the world to enjoy the city's cultural attractions. Vienna has one of the greatest collections of art treasures in Europe, showcased in such museums and art galleries as the Albertina and the Museum of Art History. Viennese opera houses, such as the State Opera House, and major theaters, such as the Burgtheater, all have worldwide reputations. The city's orchestras, including the Vienna Symphony Orchestra and the Vienna Philharmonic, have also won international fame. Throughout the summer, the city offers ballet and opera performances, concerts, and musical festivals

The Museum of Natural History opened in 1889. It maintains one of the world's largest collections of exhibits related to natural history. The building stands in the area in central Vienna known as the Ringstraasse.

The south tower of St. Stephen's Cathedral is one of the highest church towers in the world at 446 feet (136 meters). Visitors can climb 312 steps to a chamber at the top of the tower and enjoy a magnificent view of Vienna's cobbled streets, churches, and palaces.

Architectural landmarks

Vienna has also preserved most of its architectural treasures. Many of the city's historic buildings and landmarks, as well as its most fashionable shopping districts, are in the old "Inner City" section in central Vienna.

St. Stephen's Cathedral, with its high church tower, stands at the heart of the Inner City. Several blocks west is the Hofburg, a palace that combines the modern with the medieval. The royal apartments of the Hofburg, occupied by Austria's rulers for more than 600 years, are now the official residence of the president of Austria. Other buildings in the Hofburg include the Imperial Library, several museums, and the Spanish Riding School where the famous Lipizzan horses

Cakes and pastries tempt customers in a Konditorei, or Viennese pastry shop. Many people enjoy gathering at sidewalk cafes in the Inner City to drink coffee, eat cake, and watch the world go by.

are trained. Nearby are two of Vienna's most beautiful parks—the Burggarten and the Volksgarten, which is well known for its roses.

A band of streets called the Ringstrassen encircles the Inner City. Some of Vienna's most impressive public buildings line these streets, including the Museum of Art History, the City Hall, the Opera House, the Parliament Building, and the Stock Exchange. These buildings date from the second half of the 1800's.

Several important buildings, such as the Karlskirche (Church of St. Charles) and the Belvedere Palace, lie outside the Ringstrassen in the older suburban districts of the city. These structures rank among the finest existing examples of *baroque* architecture, a highly decorative style of the 1600's and 1700's.

The Schönbrunn Palace, another splendid baroque building, stands at the southwestern edge of the city. The Schönbrunn Zoo, in the palace grounds, was built in 1752 and ranks as the world's oldest zoo. A long park called the Prater lies north of the Danube. In the Prater is an amusement park with a famous Ferris wheel. Instead of seats for two people, the rim of the Ferris wheel has large enclosed cabins, each of which can carry dozens of people. The Vienna Woods line the western edge of the city.

Lipizzan horses learn to perform graceful jumping and dancing feats at the Spanish Riding School in the Hofburg. These beautiful show horses have been trained at the Viennese school for more than 400 years.

AZERBAIJAN

In 1991, Azerbaijan (AH *zuhr by JAHN*) declared itself an independent country and a member of the Commonwealth of Independent States (CIS). Azerbaijan had previously been a republic of the now-defunct Soviet Union.

Azerbaijan was under the strict control of the Soviet central government until the late 1980's, when popular opposition groups demanded greater control of the republic's own affairs. In the midst of political upheaval in the Soviet government following an attempted coup in August 1991, Azerbaijan declared its independence.

Most of Azerbaijan's people are ethnic Azerbaijanis. Armenians, Lezgins, and Russians make up the largest minority groups. Most Azerbaijanis are Shi`ite Muslims. Most Armenians are Christians. The Azerbaijani language developed from the languages of Persians and Turkic people who once inhabited the region. Today, Azerbaijani closely resembles the modern Turkish language.

Farmers in Azerbaijan grow such crops as apples, barley, cotton, grapes, tomatoes, and wheat. In the north, Azerbaijani herders graze their livestock on the pastures of the mighty Caucasus Mountains. The waters of the Caspian Sea provide large catches of carp and sturgeon, which are processed into caviar and canned products.

However, Azerbaijan's chief source of wealth is the oil that comes from rich deposits on the Abseron Peninsula. Baku, the capital of Azerbaijan, is now one of the world's major oil-producing regions. Azerbaijan's other mined products include aluminum, copper, iron, natural gas, and salt.

In addition to its oil deposits, Baku is also known for its beautiful historic buildings. Often called *the pearl of the Caspian Sea,* the city boasts a well-preserved ancient town known as the Citadel, where many architectural masterpieces from the Middle Ages have been carefully preserved. Along a maze of narrow streets and blind alleys can be found the majestic tower of Kyz Kalasy (the Maiden's Tower), the Synyk Kala minaret, and the palace of the Shirvan Shahs.

FACTS

Official name:	Azarbaycan Respublikasi (Republic of Azerbaijan)
Capital:	Baku
Terrain:	Large, flat Kur-Aras Lowland, much of it below sea level, with Great Caucasus Mountains to the north; Baku lies on Abseron Peninsula that juts into Caspian Sea
Area:	33,436 mi² (86,600 km²)
Climate:	Dry, semiarid steppe
Main rivers:	Aras, Kur
Highest elevation:	Bazardyuzyu, 14,652 ft (4,466 m)
Lowest elevation:	Caspian Sea, 92 ft (28 m) below sea level
Form of government:	Republic
Head of state:	President
Head of government:	Prime minister
Administrative areas:	59 rayonlar (rayons), 11 saharlar (cities), 1 muxtar respublika (autonomous republic)
Legislature:	Milli Mejlis (National Assembly) with 125 members serving five-year terms
Court system:	Supreme Court
Armed forces:	66,900 troops
National holiday:	Founding of the Democratic Republic of Azerbaijan - May 28 (1918)
Estimated 2010 population:	8,726,000
Population density:	261 persons per mi² (101 per km²)
Population distribution:	52% urban, 48% rural
Life expectancy in years:	Male, 66; female, 73
Doctors per 1,000 people:	3.6
Birth rate per 1,000:	18
Death rate per 1,000:	6
Infant mortality:	34 deaths per 1,000 live births
Age structure:	0-14: 24%; 15-64: 69%; 65 and over: 7%
Internet users per 100 people:	11
Internet code:	.az
Languages spoken:	Azerbaijani (official), Lezgi, Russian, Armenian
Religions:	Muslim 93.4%, Russian Orthodox 2.5%, Armenian Orthodox 2.3%, other 1.8%
Currency:	Manat
Gross domestic product (GDP) in 2008:	$46.38 billion U.S.
Real annual growth rate (2008):	11.6%
GDP per capita (2008):	$5,388 U.S.
Goods exported:	Cotton, food, machinery, oil and gas
Goods imported:	Food, machinery and equipment, metals, oil products
Trading partners:	Iran, Italy, Russia, Turkey, United Kingdom

Azerbaijan lies on the western shore of the Caspian Sea. The lowlands of the River Kur and its tributary, the Aras, cover most of the republic.

Azerbaijan's capital city, Baku, which lies 40 feet (12 meters) below sea level on the Caspian Sea, is the center of a great oil industry.

Azerbaijan's location between the Caucasus Mountains and the Caspian Sea has made it a strategically important area since ancient times. Over the years, Mongols, Persians, and Turks took control of Azerbaijan. Then, in the late 1820's, after a successful war with Persia, Russia made Azerbaijan part of its empire. In 1920, Communists gained power in Azerbaijan. In 1922, Azerbaijan joined with Soviet Armenia and Georgia to form the Transcaucasian Federation. This federation was one of the four republics that formed the Soviet Union later that year.

In 1923, the mainly Armenian district of Nagorno-Karabakh became an autonomous region within Azerbaijan. The region of Nakhichevan was incorporated into Azerbaijan in 1924.

When Nagorno-Karabakh voted to secede from Azerbaijan in 1988, the long-standing hostility between the Muslim Azerbaijanis and their Christian Armenian neighbors erupted into bloody violence. Political turmoil continued in 1993. That year, former Communist Heydar Aliyev was appointed acting president after elected president Abulfaz Elcibey was forced from office by a military revolt. Later, Aliyev was elected president.

In May 1994, Aliyev signed a truce with Armenia even though it left Armenia in control of 20 percent of Azerbaijan. In October, special police mounted an abortive coup against Aliyev. In 2003, Aliyev did not run for re-election due to ill health. Voters elected his son Ilham to succeed him as president. Heydar Aliyev died later that year. Ilham Aliyev was reelected in 2008.

THE AZORES AND MADEIRAS

For a farmer and his son horseback is still the best way to get around the steep hills of the Azores.

In addition to its mainland territory, the nation of Portugal also includes two island groups in the Atlantic Ocean, the Azores *(AY zohrz* or *uh ZOHRZ)* and the Madeiras *(muh DEER uhz)*. The Azores lie about 800 miles (1,300 kilometers) off the west coast of Portugal. The Madeiras are situated about 350 miles (560 kilometers) off the northwest coast of Morocco. Both are *autonomous* (self-governing) regions of Portugal.

The Azores

The Azores are actually the peaks of a huge undersea volcanic mountain range that extends down the Mid-Atlantic Ridge from Iceland almost to Antarctica. The nine islands that make up the Azores cover 897 square miles (2,322 square kilometers) and have a 320-mile (515-kilometer) coastline. About 248,000 people live in the Azores.

Navigator Gonzalo Cabral claimed the Azores for Portugal in 1431. At the time, the islands were uninhabited, but Portuguese people from the mainland soon began to settle on the islands. They were later joined by Flemish, Breton, and Spanish immigrants. The islands were never a colony; they were always considered part of Portugal.

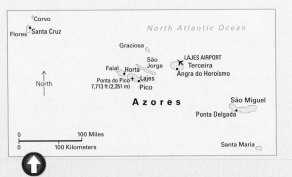

The Azores have been part of Portugal since they were claimed in 1431 by Portuguese explorers. The seafarers named the island "Azores" after the hawks that constantly circle the islands. They thought the birds were vultures.

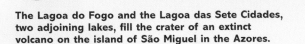

The Lagoa do Fogo and the Lagoa das Sete Cidades, two adjoining lakes, fill the crater of an extinct volcano on the island of São Miguel in the Azores.

The Azores played an important role in World War II (1939–1945). Because of the islands' strategic location, the United Kingdom used them as a naval base in their fight against Nazi submarines. Although Portugal remained neutral, an ancient treaty allowed the United Kingdom to use the islands in time of war. Today, the United States keeps military installations there.

Hot springs, geysers, and huge craters are reminders of the islands' volcanic beginnings, and the forces of nature can still be felt during occasional earthquakes. In some areas, lush green vegetation extends right down to the sea, creating a vast blanket of exotic flowers. The climate is cool and humid throughout the year.

Farmers grow corn, grapes, and citrus fruits, and cattle graze on stretches of grassland. Fishing crews catch tuna. Manufacturing on the islands is modest, but pottery, leather goods, and other handicrafts are produced to meet the demands of the tourist industry.

The Madeira Islands

The ancient Romans named these beautiful islands *Pur-puriarae* (Purple Islands). The Portuguese, who arrived in 1419, gave them their present name, which means *wood*, because of the extensive forests they found there. These are the Madeiras—born of volcanic lava, home of the world-famous Madeira wine and, some say, a breathtaking glimpse of paradise.

The Madeiras consist of the main island of Madeira, its smaller neighbor Porto Santo, and two uninhabited groups of islets, the Desertas and the Selvagens. The islands cover a total area of 309 square miles (801 square kilometers). Like the Azores, the Madeiras are the peaks of undersea volcanic mountain ranges.

The island of Madeira is the largest and most important island in the chain. Madeira rises to its highest point at Pico Ruivo, which has an elevation of 6,104 feet (1,860 meters). Deep valleys cut through its mountainous landscape, and steep cliffs rise sharply out of the sea on the northeast side. The island's settlements and farms are built on terraces, which are covered with exotic flowers and trees.

After the Portuguese arrived in 1419, Madeira became a regular port of call for seafarers. Ships from all over Europe, the Far East, and the New World stopped there, and many left specimens of exotic plants. The island's warm, sunny climate and rich volcanic soil soon turned Madeira into a garden of subtropical fruits, vegetables, and flowers.

Most of the Madeiras' population, which numbers about 250,000, make their living in agriculture or tourism. Chief crops include bananas, corn, mangoes, oranges, pomegranates, sugar cane, vegetables, and wine grapes. Because rain falls only in winter, water for the crops is stored and distributed by stone aqueducts called *levadas*.

Madeira has long been one of Portugal's most popular tourist stops. Funchal, the capital of the Madeiras as well as the largest city and chief seaport, is the islands' center for tourism. Many people, including children, work in the islands' cottage industries, making embroidery, lace, and willow wicker baskets.

Volcanic formations, framed by the shimmering blue waters of the Atlantic, form a dramatic background as fishing crews work in Madeira's coastal waters. Black scabbard—fish caught at a depth of 6,600 feet (2,000 meters)—is a local delicacy.

Portuguese explorers sailed to the Madeiras in 1419. The islands are best known for their fine Madeira wines. Wine-making is the principal industry.

THE BAHAMAS

In 1492, Christopher Columbus first landed in America at what is now San Salvador Island in the Bahamas. Today, this independent nation consists of nearly 700 islands and about 2,300 rocky islets and reefs near the coasts of Florida and Cuba.

Most of the islands are limestone with only a thin layer of infertile, stony soil. Many of them are partly covered with pine forests. People live on only about 20 of the islands, but the islands' beauty and mild climate attract many tourists every year.

People and history

People of African descent make up about four-fifths of the population of the Bahamas. The rest are mainly whites or *mulattoes* (people of mixed African and European ancestry).

Many black Bahamians are descendants of Africans brought to the islands by British settlers to work as slaves. Many of these British settlers came from the United States after the Revolutionary War ended in 1783.

The British began settlements in the Bahamas as far back as the 1600's. The islands had been claimed for Spain in 1492, but the Spaniards never settled the islands. Instead, they enslaved the Lucayo Indians who lived there and took many of them to work in gold mines on nearby islands.

Spain began to attack the British settlements in the late 1600's. Pirates, who used the coves and islets as bases for their raids, also attacked the settlements. The Bahamas became a British colony in 1717, and Spain gave up its claim to the islands in 1783.

In the mid-1800's, the Bahamas prospered from shipping and trade. During the American Civil War, the Bahamas were used as a base for ships breaking the Union blockade of Southern ports. After the war, the Bahamian economy declined, but prosperity returned about 100 years later, when large numbers of tourists began visiting the islands.

In 1964, the United Kingdom granted the Bahamas internal self-government, and in 1967 the Progressive Liberal Party, made up largely of Bahamians of African

FACTS

Official name:	Commonwealth of the Bahamas
Capital:	Nassau
Terrain:	Long, flat coral formations with some low rounded hills
Area:	5,382 mi² (13,940 km²)
Climate:	Tropical marine; moderated by warm waters of Gulf Stream
Main rivers:	N/A
Highest elevation:	Mount Alvernia on Cat Island, 206 ft (63 m)
Lowest elevation:	Atlantic Ocean, sea level
Form of government:	Constitutional monarchy
Head of state:	British monarch, represented by governor general
Head of government:	Prime minister
Administrative areas:	21 districts
Legislature:	Parliament consisting of the Senate with 16 members serving five-year terms and the House of Assembly with 40 members
Court system:	Supreme Court; Court of Appeal; magistrate courts
Armed forces:	860 troops
National holiday	Independence Day - July 10 (1973)
Estimated 2010 population:	342,000
Population density:	64 persons per mi² (25 per km²)
Population distribution:	84% urban, 16% rural
Life expectancy in years:	Male, 66; female, 73
Doctors per 1,000 people:	1.1
Birth rate per 1,000:	17
Death rate per 1,000:	8
Infant mortality:	19 deaths per 1,000 live births
Age structure:	0-14: 27%; 15-64: 67%; 65 and over: 6%
Internet users per 100 people:	42
Internet code:	.bs
Languages spoken:	English (official), Creole (among Haitian immigrants)
Religions:	Baptist 35.4%, Anglican 15.1%, Roman Catholic 13.5%, Pentecostal 8.1%, Church of God 4.8%, other Christian 19.4%, other 3.7%
Currency:	Bahamian dollar
Gross domestic product (GDP) in 2008:	$6.94 billion U.S.
Real annual growth rate (2008):	1.5%
GDP per capita (2008):	$20,640 U.S.
Goods exported:	Chemicals, food, mineral products, rum
Goods imported:	Chemicals, food, machinery, mineral products, transportation equipment
Trading partners:	Germany, Japan, Singapore, South Korea, Spain, United States

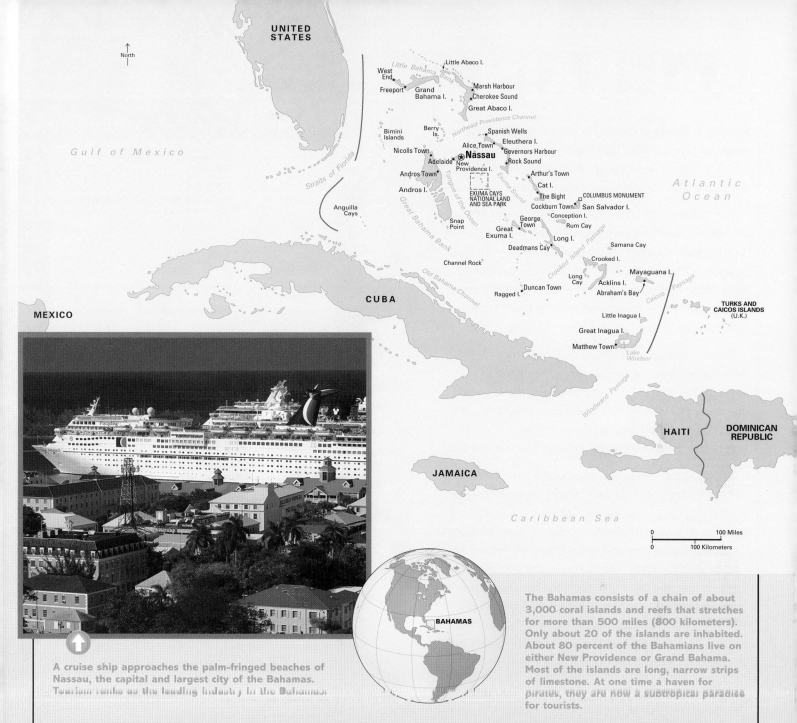

The Bahamas consists of a chain of about 3,000 coral islands and reefs that stretches for more than 500 miles (800 kilometers). Only about 20 of the islands are inhabited. About 80 percent of the Bahamians live on either New Providence or Grand Bahama. Most of the islands are long, narrow strips of limestone. At one time a haven for pirates, they are now a subtropical paradise for tourists.

A cruise ship approaches the palm-fringed beaches of Nassau, the capital and largest city of the Bahamas. Tourism ranks as the leading industry in the Bahamas.

descent, won control of the government. For the first time, the black majority was in power. The government then worked for full independence, which was achieved on July 10, 1973.

Government and economy

The monarch of the United Kingdom, represented by a governor general, is the official head of state of the Commonwealth of the Bahamas. Voters elect the members of the House of Assembly, one house of the legislature. The head of the party that wins the most Assembly seats becomes prime minister. Members of the second house of the legislature, the Senate, are appointed.

Tourism is the leading economic activity of the Bahamas. Many Bahamians work in hotels or other tourist-related businesses. Less than 2 percent of the people farm the land. Farmers grow bananas, citrus fruits, cucumbers, pineapples, tomatoes, and other crops. Fishermen catch crawfish and other seafood for domestic consumption and for export.

Food processing ranks as a major industry. Foreign corporations run businesses in the Bahamas, and the country has branches of many foreign banks.

BAHRAIN

Bahrain is an island country in the Persian Gulf. It is made up of more than 30 islands, including the largest, which is also named Bahrain. Although surrounded by water, the islands have a dry desert climate.

For hundreds of years, Bahrain was a center of trade and communications in the Persian Gulf region. Dilmun, a prosperous trading civilization, occupied the islands about 4,000 years ago. In the 1700's, al-Khalifah Arabs from Saudi Arabia took control of Bahrain and have ruled it ever since, though the country was a protectorate of the United Kingdom from 1861 to 1971.

Bahrain was an underdeveloped nation until 1932, when petroleum was discovered on the island of Bahrain. The country now enjoys one of the highest standards of living in the Persian Gulf area. It has one of the highest literacy rates in the region, and education is free. The government provides free medical care.

In 2001, Bahraini voters approved a national charter to reform their country's government. The reforms went into effect in 2002, changing Bahrain from an *emirate*, ruled by an *emir* with absolute power, to a constitutional monarchy, with a king and a two-house legislature. The people elect the members of one house, and the king appoints the members of the other. The country previously had a national assembly elected by the people, but it was disbanded by the emir in 1975. In the mid-1990's, demands for the restoration of parliament led to violent antigovernment riots and bombings.

In 2011, antigovernment protests erupted again. The protesters sought greater democratic reform and the removal of the king, Sheik Hamad bin Isa Al-Khalifa. In response, in 2012, the king announced constitutional reforms that somewhat increased the elected parliament's power. Nevertheless, unrest continued.

About 80 percent of Bahrain's people are Arabs. Large groups of Indians, Iranians, and Pakistanis also live in the country. Almost all the people are Muslims, and Islam is the national religion. Arabic is the official language, though Farsi—the language of Iran—and English are also spoken. Many Bahrainis, especially younger people, wear clothes reflecting Western influence, but others still wear traditional Arab dress.

FACTS

Official name:	Mamlakat al Bahrayn (Kingdom of Bahrain)
Capital:	Manama
Terrain:	Mostly low desert plain rising gently to low central escarpment
Area:	277 mi² (718 km²)
Climate:	Arid; mild, pleasant winters; very hot, humid summers
Main rivers:	N/A
Highest elevation:	Jabal ad Dukhan, 443 ft (135 m)
Lowest elevation:	Persian Gulf, sea level
Form of government:	Constitutional monarchy
Head of state:	King
Head of government:	Prime minister
Administrative areas:	12 manatiq (municipalities)
Legislature:	Parliament consisting of the Shura Council with 40 members appointed by the king and the House of Deputies with 40 members serving four-year terms
Court system:	High Civil Appeals Court
Armed forces:	8,200 troops
National holiday:	National Day - December 16 (1971)
Estimated 2010 population:	794,000
Population density:	2,866 persons per mi² (1,106 per km²)
Population distribution:	95% urban, 5% rural
Life expectancy in years:	Male, 73; female, 77
Doctors per 1,000 people:	2.7
Birth rate per 1,000:	19
Death rate per 1,000:	4
Infant mortality:	12 deaths per 1,000 live births
Age structure:	0-14: 27%; 15-64: 70%; 65 and over: 3%
Internet users per 100 people:	52
Internet code:	.bh
Languages spoken:	Arabic, English, Farsi, Urdu
Religions:	Muslim 81.2%, Christian 9%, other 9.8%
Currency:	Bahraini dollar
Gross domestic product (GDP) in 2008:	$18.92 billion U.S.
Real annual growth rate (2008):	6.1%
GDP per capita (2008):	$26,198 U.S.
Goods exported:	Mostly: petroleum and petroleum products Also: aluminum, chemicals
Goods imported:	Crude oil, food, industrial machinery, motor vehicles
Trading partners:	Germany, Japan, Saudi Arabia, United States

Most of Bahrain's people live in towns in the northern part of the island of Bahrain. The majority live in houses or apartments, but some villagers build thatched huts. Freshwater springs provide ample drinking water on the northern coast of Bahrain. Farmers also use this water to irrigate their land.

Northern Bahrain receives most of the little rain that falls—about 3 inches (8 centimeters) a year, mainly during the winter. Summers are hot and humid. Bahrain has excellent electric service systems, so refrigerators and air conditioners are common.

The electricity, free medical care, and other services Bahrainis enjoy are due mainly to oil profits. Bahrain actually has only a small supply of petroleum, but its large oil refinery, on the island of Sitrah, processes all of Bahrain's crude petroleum, as well as much of the oil that comes from Saudi Arabia via pipeline.

While Bahrain's economy depends largely on the petroleum industry, the government has also established programs designed to develop commerce, construction, fishing, manufacturing, and transportation. In addition, Bahrain ranks as a major banking and financial center for the Persian Gulf region. Modern warehouse and port facilities help make it a major trading center as well.

Ship repairing is also an important industry in Bahrain. Factories produce aluminum, ammonia, liquid natural gas, and methanol. Agriculture forms a small part of Bahrain's economy. Farmers grow dates, tomatoes, and other fruits and vegetables on irrigated land. They also raise cattle, goats, and poultry.

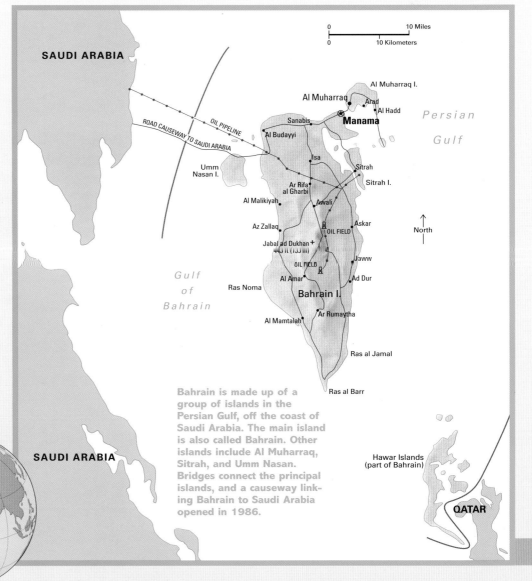

Bahrain is made up of a group of islands in the Persian Gulf, off the coast of Saudi Arabia. The main island is also called Bahrain. Other islands include Al Muharraq, Sitrah, and Umm Nasan. Bridges connect the principal islands, and a causeway linking Bahrain to Saudi Arabia opened in 1986.

Bangladesh, an independent nation since 1971, is almost completely surrounded on three sides by the northeast part of India. Myanmar touches Bangladesh's southwest coast, and its south coast opens to the Bay of Bengal, which is the northern part of the Indian Ocean.

Rivers and rich soil

Bangladesh has some of the richest, most fertile soil in the world. Most of the country consists of an *alluvial plain,* which is land formed from soil left by rivers.

Three major rivers—the Brahmaputra, the Ganges, and the Meghna—flow through Bangladesh and into the Bay of Bengal. Another of the country's rivers, the Padma, is a branch of the Ganges. The Padma River begins where the Ganges meets the Jamuna River. It then flows 78 miles (126 kilometers) to join the Meghna River at Chandpur.

During the rainy season, the rivers of Bangladesh overflow and deposit sediment along their banks. The built-up soil deposits form the Ganges Delta.

Rice, the country's main source of food, and *jute,* a plant whose fibers are made into string or woven into cloth, thrive in the wet delta region. They are two of the country's most important crops. Bangladesh, in fact, is one of the world's leading rice producers.

The rivers that provide Bangladesh's rich soil begin high in the Himalaya. However, the clearing of forests on the mountains causes much of the region's soil to wash down the slopes. As a result, sediment builds up in the riverbeds. When the snow melts in the mountains and the monsoon rains arrive, the rivers cannot contain the extra water. The rivers then overflow and flood the surrounding countryside.

The struggle for independence

Bangladesh was formerly part of Pakistan, which was established in 1947 when India became independent. Because of the fighting between India's Hindus and Muslims under British rule, Indian and British leaders decided to divide India into two countries. India became the Hindu nation, and Pakistan became the Muslim nation.

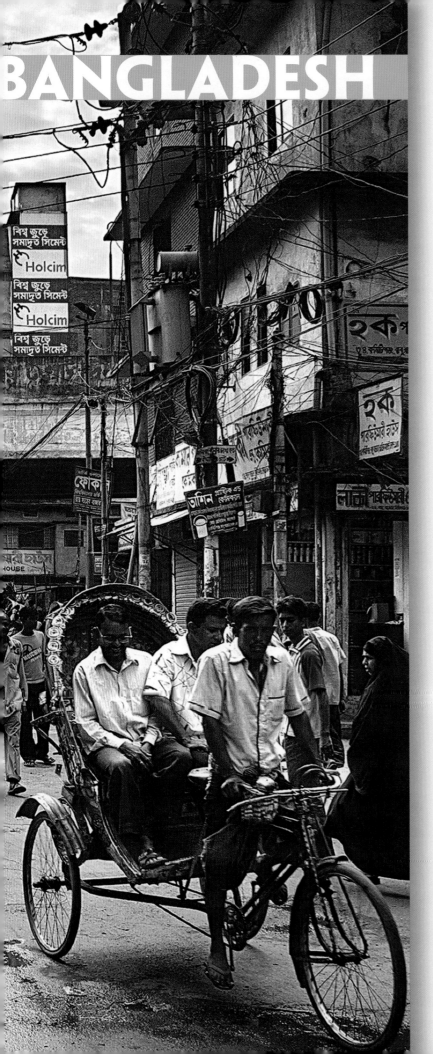

BANGLADESH

Pakistan was further divided into West Pakistan, on India's northwest border, and East Pakistan, on India's northeast border. Unfortunately, the two areas had little in common except their religion.

Through the years, East Pakistanis became dissatisfied with the government of Pakistan. In 1971, fighting broke out. That same year, East Pakistan declared its independence from West Pakistan and became the nation of Bangladesh.

Early beginnings

The country that is now often called *the poorest of the poor* was once a busy and successful commercial center.

Bangladesh is part of the region known as Bengal, which also includes the Indian state of West Bengal. In the 1500's, European traders following the traditional Indian Ocean routes came to Bengal. There, they found a rich land dotted with many small commercial centers. At the heart of these centers was a busy handloom weaving industry that produced high-quality textiles.

The Europeans were interested in Bengal's many rivers. Because they led deep into the country's interior, the rivers were ideal for transporting goods.

By the early 1700's, the Mogul Empire, which ruled Bengal at the time, had grown steadily weaker. Soon, the British East India Company became the most powerful force in Bengal.

The British then decided to move their center of operations to Calcutta in West Bengal. This move proved to be a fateful turn of events for the region as commercial interests and investment money moved steadily westward.

At the same time, the Industrial Revolution in Europe brought mechanization to the British textile industry. Textile exports from Bengal were discouraged, and British-made goods flooded the region.

Bengal's craft industries were completely ruined, and its trade and commerce collapsed. As a result, the region became a purely agricultural society during the 1800's, with few opportunities for development.

119

BANGLADESH TODAY

Bangladesh is a poor, underdeveloped country. Throughout its short history, it has seen more than its share of civil war, natural disaster, and poverty. A civil war brought Bangladesh into being, and the conflict left the nation with serious economic, social, and political problems. Many poor Bangladeshis cannot afford a sufficient supply of food. Cyclones, tornadoes, tidal waves, and floods occur almost every year. These disasters cause much death and destruction.

Strong beliefs and hardships

The short and troubled history of Bangladesh has its beginnings when the country was still known as East Pakistan. In 1970, a cyclone and tidal wave struck East Pakistan, killing about 266,000 people. Many of the survivors believed that the West Pakistan government held back on relief shipments. Tensions between East and West Pakistan grew.

In a December 1970 election to select a new assembly, the Awami League, a political party led by East Pakistan's Sheik Mujibur Rahman, won a majority of the seats. The Awami League strongly supported increased self-government for East Pakistan.

When the president of Pakistan postponed the first meeting of the assembly in March 1971, East Pakistan protested. West Pakistan troops were brought in to put down the protest, and Sheik Mujibur Rahman (known as Sheik Mujib) was imprisoned in West Pakistan. On March 26, 1971, Bangladesh declared itself an independent nation.

The civil war that followed lasted until December 16. With the help of Indian troops, Bangladesh defeated West Pakistan, but the effects of the civil war were devastating to the new nation. Even as the government set up programs to rebuild the country, floods and food shortages brought more troubles.

Mujib was released from prison and returned to Bangladesh, where he became the nation's first prime minister in 1972. In 1975, after changing the constitution to give all executive power to the president,

FACTS

Official name:	People's Republic of Bangladesh
Capital:	Dhaka
Terrain:	Mostly flat alluvial plain; hilly in southeast
Area:	55,598 mi² (143,998 km²)
Climate:	Tropical; cool, dry winter (October to March); hot, humid summer (March to June); cool, rainy monsoon (June to October)
Main rivers:	Brahmaputra, Ganges, Meghna, Jamuna
Highest elevation:	Mount Keokradong, 4,034 ft (1,230 m)
Lowest elevation:	Indian Ocean, sea level
Form of government:	Parliamentary democracy
Head of state:	President
Head of government:	Prime minister
Administrative areas:	6 divisions
Legislature:	Jatiya Sangsad (National Parliament) with 300 members serving five-year terms
Court system:	Supreme Court
Armed forces:	157,100 troops
National holiday:	Independence Day - March 26 (1971)
Estimated 2010 population:	161,315,000
Population density:	2,901 persons per mi² (1,120 per km²)
Population distribution:	74% rural, 26% urban
Life expectancy in years:	Male, 60; female, 63
Doctors per 1,000 people:	0.3
Birth rate per 1,000:	25
Death rate per 1,000:	8
Infant mortality:	52 deaths per 1,000 live births
Age structure:	0-14: 34%; 15-64: 62%; 65 and over: 4%
Internet users per 100 people:	0.3
Internet code:	.bd
Languages spoken:	Bengali (official), English
Religions:	Muslim 89.6%, Hindu 9.3%, other 1.1%
Currency:	Taka
Gross domestic product (GDP) in 2008:	$81.32 billion U.S.
Real annual growth rate (2008):	4.9%
GDP per capita (2008):	$542 U.S.
Goods exported:	Mostly clothing Also fish products, jute and jute goods, leather
Goods imported:	Building materials, chemicals, food, machinery, petroleum, textiles
Trading partners:	China, Germany, India, United Kingdom, United States

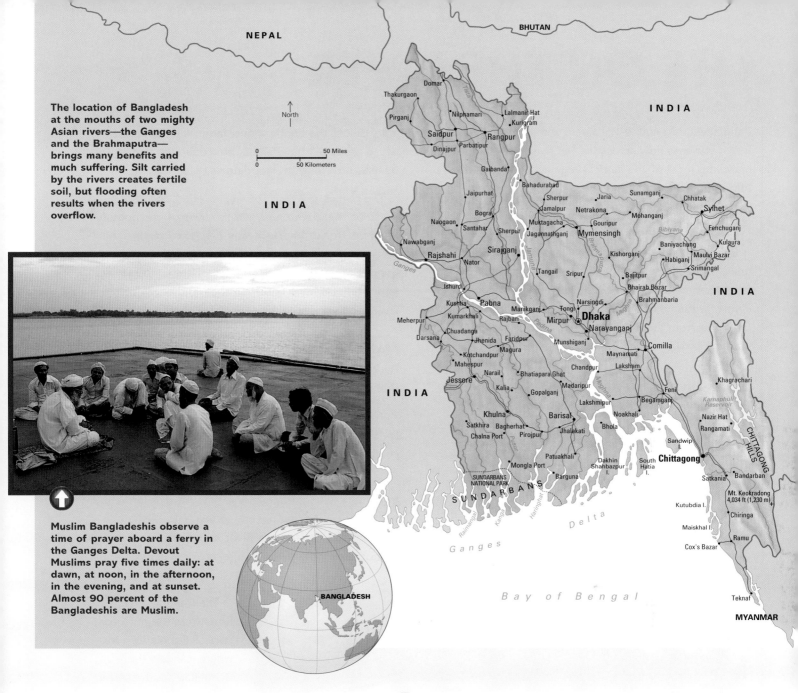

The location of Bangladesh at the mouths of two mighty Asian rivers—the Ganges and the Brahmaputra—brings many benefits and much suffering. Silt carried by the rivers creates fertile soil, but flooding often results when the rivers overflow.

Muslim Bangladeshis observe a time of prayer aboard a ferry in the Ganges Delta. Devout Muslims pray five times daily: at dawn, at noon, in the afternoon, in the evening, and at sunset. Almost 90 percent of the Bangladeshis are Muslim.

Mujib resigned as prime minister and took office as president. Later that year, military leaders killed Mujib and instituted martial law. Elections were held in 1978, and martial law ended in 1979, though political unrest continued. In 1991, the constitution was again amended, and executive power was returned to the prime minister. The position of president became largely ceremonial.

Throughout the 1990's and the first decade of the 2000's, Bangladesh's government remained weak, with frequent boycotts by opposition forces. In 2009, Sheik Hasina Wajed, Mujib's daughter, was elected prime minister in elections that were considered democratic.

Economy

The country's warm, humid climate and fertile soil are ideal for farming. However, many Bangladeshi farmers have little or no land of their own and use the same simple tools that their ancestors used. Jute is the chief export crop of Bangladesh. Farmers also grow rice, sugar cane, tea, tobacco, and wheat.

Bangladesh ranks as one of the poorest nations of the world. However, natural gas and petroleum have been discovered in the northeast and in the coastal waters. Industry is growing, especially the manufacture of clothing and textiles. The country's economic future depends largely on its ability to solve the growing problem of overpopulation.

LIFE ON THE RIVER

Bangladesh is a country of rivers and streams. Countless waterways flow across the flat river plain that makes up most of the land. Three major Asian rivers—the Brahmaputra, the Ganges, and the Meghna—unite in Bangladesh to form the great Ganges Delta at the Bay of Bengal.

Bangladesh depends on the rivers for its very existence. Most of the country, which lies only about 50 feet (15 meters) above sea level, would be permanently flooded if the river system did not carry away the water from the monsoon rains and the melting Himalayan snows.

A life-giving force

For the people of Bangladesh, the rivers are both a life-giving force and an instrument of destruction. The soil deposited along their fertile banks creates some of the richest farmland in the world, but when the rivers overflow, the flooding brings death and disaster.

Life in Bangladesh revolves around the rivers. Villagers watch the changing flow of the waters and build their houses on whatever high ground is available. Before the floods arrive, farmers try to guess which areas will be affected, how deep the floods will be, and how long the flooding will last.

The names given to the rivers reflect the importance of these waterways in the lives of the people. *Khlamati*, for example, means *fulfiller of desires. Kirtinasha* means *destroyer of achievements*. River themes are common in the tales of the *alapanis* (story-tellers), who entertain their audiences with a variety of ancient tales.

Major rivers

Bangladesh's major rivers include the Ganges, the Brahmaputra, the Jamuna, and the Meghna.

The 1,560-mile-long (2,510-kilometer-long) Ganges River extends from the western border with India to its confluence with the Jamuna River.

The Brahmaputra *(son* of *Brahma)* River flows down from India. Where it meets the Tista River, it forms the Jamuna River. No permanent settlements exist along the banks of the Jamuna River. Its violent floods, which do not allow the soil to mature, create a landscape of *chars* (seasonal islands).

The Meghna River flows from the northeast part of Bangladesh to where it meets the Ganges in south-central Bangladesh.

Fishing boats set off down a river. The nation's waterways contain large supplies of fish, which is an important food source. However, a traditional disrespect for fishermen has slowed the development of a fishing industry.

Houses are built on stilts as protection against rising floodwaters. Annual monsoon flooding causes great loss of human life, as well as damage to property and communication systems. The government has developed major flood-control projects.

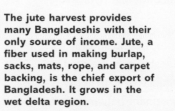

The jute harvest provides many Bangladeshis with their only source of income. Jute, a fiber used in making burlap, sacks, mats, rope, and carpet backing, is the chief export of Bangladesh. It grows in the wet delta region.

PEOPLE

The ancestors of present-day Bangladeshis probably came from Myanmar, Tibet, and northern India thousands of years ago. More than 95 percent of Bangladesh's population belong to the Bengali ethnic group. Their national language is Bengali, which has a rich cultural heritage in literature, music, and poetry.

Tribal groups form the second largest ethnic group in Bangladesh. The four largest tribes are the Chakmas, the Marmas, the Mros, and the Tipperas. They live primarily in the Chittagong Hills of southeastern Bangladesh. The culture of the tribal groups differs greatly from that of other Bangladeshis. Most follow the Buddhist religion and speak Tibeto-Burman languages.

Another ethnic group, the Biharis, includes the Urdu-speaking, non-Bengali Muslim refugees from Bihar and other parts of northern India. The Biharis were once the upper class of Bengali society. Many held jobs on the railroads and in heavy industry.

The Biharis did not want East Pakistan to separate from West Pakistan in 1971 because they felt financially threatened by the separation. After Bangladesh became independent, hundreds of thousands of Biharis returned to West Pakistan. Those who remain live in their own small communities. They are often persecuted for their loyalty to Pakistan.

Heritage and religion

Bangladeshis take great pride in their Bengali heritage. Their national language, Bengali, is particularly important to them, and the religion of Islam provides an important foundation for their daily life.

Islam came to Bengal long after it was established in Pakistan. The Muslims conquered Pakistan in A.D. 711. Islam did not reach the Bengalis until 500 years later, when Turkish Muslims extended their control into the eastern region of Bengal. Before that, Hindu or Buddhist dynasties had ruled Bengal.

Today, many strict Muslims do not accept Bangladeshis as true Muslims. They believe that Bangladeshis owe their faith to conversion rather than natural inheritance. This belief was one of the major reasons for the disagreement between East and West Pakistan.

Villagers honor the river in a ceremony along the banks of the Kali River. The rivers of Bangladesh provide fish for food and for export. Rivers are also the country's chief transportation routes.

A Muslim wedding party leaves the ceremony by bus. Men in Muslim families have far more freedom than women. Muslim women have few activities outside the home, and many cover their heads with veils in the presence of strangers.

Waterways take the place of roads in Bangladesh, where countless rivers and streams flow through the country. Large passenger and cargo ships provide transportation between the cities. Smaller boats operate between villages.

The fertile soil of Bangladesh yields two—and often three—harvests a year, but land that could grow crops is now used for living space. The rapid population growth has kept most Bangladeshis poor. Many people have moved from rural villages to urban areas in an effort to improve their lives, but few jobs are available.

Population growth

Bangladesh is one of the world's most densely populated countries. In 2010, the population of Bangladesh was around 160 million—about 30 million more than in 2000. Bangladesh does not have the resources to feed and educate its huge population. Many Bangladeshis do not have enough to eat, and most adults cannot read and write.

Because of the rapid population growth, poverty and overcrowding are serious problems in both urban and rural areas. Rural villagers live in one- or two-room huts made of bamboo. In urban slums, shelters are built of cardboard, scraps of wood, or sticks.

Way of life

Bangladesh ranks as one of the poorest nations in the world. But in spite of their poverty and hardships, the men enjoy gathering in cafes and marketplaces, while women visit each other's homes. Villagers are often entertained by story-tellers, and they also enjoy singing and listening to folk ballads. Craft workers carry on the traditions of embroidery, weaving, pottery, and other decorative arts.

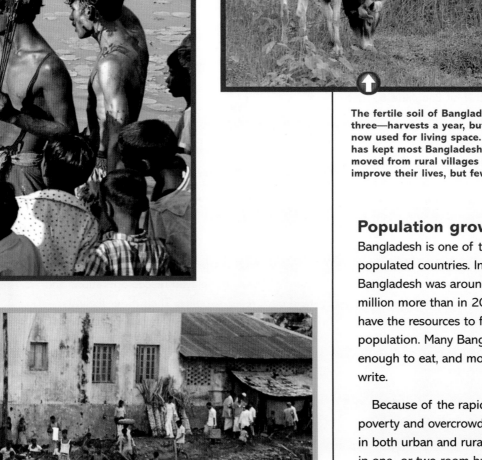

A COUNTRY AT RISK

Bangladesh is both a beneficiary and a victim of its climate and geography. The country receives a great deal of rainfall, brought mostly by the monsoon between mid-May and October. The far northeast region gets the most rain—as much as 250 inches (635 centimeters) a year.

Too much rainfall during the monsoon season causes large-scale flooding, while too little rainfall brings drought. Cyclones, which sweep into the country from the Bay of Bengal, are also a constant threat. Numerous severe cyclones have hit Bangladesh. These storms usually strike at the end of the monsoon season and may also be accompanied by tidal waves.

These natural disasters are a sad fact of life in Bangladesh. Floods, cyclones, and tidal waves have had an important effect on the history of the country. It was, in fact, a cyclone that helped bring the nation of Bangladesh into being.

In November 1970, a cyclone rose in the Bay of Bengal, and a tidal wave swept over the countryside of what was then East Pakistan. It killed about 266,000 people and destroyed many villages.

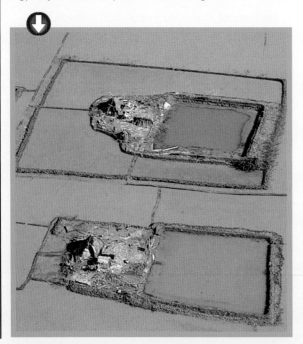

The monsoon rains that bring many benefits to Bangladesh can also result in major flooding, crop destruction, and food shortages.

Cyclones usually strike at the end of the monsoon season. Contaminated floodwaters bring even more problems. Disease spreads quickly, and food supplies are destroyed. Floods also isolate villages from rescue operations.

The victims of the disaster felt that the government of Pakistan delayed rescue operations and held back relief supplies. Their bitter feelings helped trigger the civil war that led to the independence of Bangladesh.

After the war, the new government set up many programs to help rebuild the country, and a huge international relief effort was coordinated by the United Nations. In mid-1974, just as the programs were beginning to make progress, disaster struck again. The worst floods in decades swept through Bangladesh. By September, food shortages brought famine, and tens of thousands of people died.

Weather-related problems continued throughout the 1970's and 1980's. A devastating cyclone struck the coastal areas in May 1985. The storm began to develop on May 22 in the Bay of Bengal and swept onto the coast in the early hours of

Bangladeshis swarm across a new canal. Water-control projects are designed to reduce the effects of floodwaters on farmland and other property. With so many mouths to feed, protecting the food supply is the government's most important task.

Possible sea-level rise impact in Bangladesh

Area of highest population density

Land areas submerged from a sea-level rise of 5 ft (1.5 m)

Saidpur
Rangpur
Mymensingh
Rajshahi
Sirajganj
Pabna
Mirpur · Dhaka
Narayanganj
Comilla
Jessore
Khulna · Barisal
Chittagong

North

Bay of Bengal

0   50 Miles
0   50 Kilometers

Some scientists believe that global warming will significantly raise ocean levels over the next few decades. This map shows areas of Bangladesh would be permanently flooded if sea levels rose by 5 feet (1.5 meter). Permanently flooded areas would mean even less space for the people to live and grow food—a disaster for this overpopulated country. Some scientists also predict that global warming will change rainfall patterns and bring more cyclones.

May 24. The cyclone's winds hit the coast at more than 80 miles (130 kilometers) per hour and created tidal waves 13 feet (4 meters) high.

Many Bangladeshis were unprepared for this particular cyclone. The bamboo and jute dwellings in which the villagers took shelter were no match for the storm's tremendous strength. Thousands of people were swept away by floods and tidal waves.

The 1985 cyclone killed 11,000 people, damaged more than 94,000 houses, and killed much livestock. But it was not to be the end of weather troubles for Bangladesh. The summer of 1988 brought the worst floods in the nation's history.

Caused by heavy runoff of monsoon rains in the Himalaya, the floods killed about 1,600 people, and an additional 500 died from diseases resulting from the floods. The damage to land and property was far worse than the 1985 disaster. In some districts, the entire population was left homeless.

About 10 million acres (4 million hectares) of crops were completely destroyed, and an additional 7.4 million acres (3 million hectares) were partly destroyed. Also, many railroad bridges and tracks, roads, schools, and other buildings were completely or partly destroyed.

Only three months later, a killer cyclone struck southeastern Bangladesh. About 600 people were killed, and more than 100 ships sank or ran aground as 10- to 15-foot (3- to 4.6-meter) waves hit the coastal areas.

It is clear that Bangladesh will always suffer from weather-related problems. Damage from tidal waves and cyclones is unavoidable, but if people get some advance warning, they can take shelter in safe public buildings.

On average, a catastrophe that destroys 10 percent or more of the nation's food supply is likely to occur every two or three years. This makes it even more difficult for Bangladesh to provide for its people.

BARBADOS

Of all the islands in the West Indies, Barbados *(bahr BAY dohz)* is the most British in character. Like London itself, the capital city of Bridgetown has a Trafalgar Square, as well as a Nelson's Column even older than its better-known counterpart across the sea. The distinct British flavor of Barbados, along with its beautiful landscape, gives the island a unique and special charm.

Most of Barbados is flat, with a high, rugged region lying in the middle of the northeast coast. The land descends from this region to a lowland plateau that stretches to the coast all around the island. Fine sandy beaches extend along the west and southwest coasts.

Little of the island's original vegetation remains, and most of Barbados' land is covered by vast fields of sugar cane. But at Turner's Hall Woods and Welch-man Hall Gully, in the interior of the island, small patches of the tropical rain forest that once covered Barbados still survive. Many vine-covered trees—including locust, mastic, Spanish oak, and kapok—tower over the land, and the rain forest provides shelter for such introduced animals as wild green monkeys, hares, mongooses, and colorful tropical birds.

History

British settlers arrived in Barbados in 1625, and the first permanent British settlement was established in 1627. In 1639, the landowners of Barbados elected a House of Assembly. The new colony prospered, and many British families settled on the island in the 1700's and 1800's.

In 1958, Barbados joined the West Indies Federation, a union of British islands in the West Indies. But

Passengers keep cool in an open-sided bus in Bridgetown, the capital of Barbados. Many of the graceful buildings in this busy port date from the time the British ruled the island.

FACTS

Official name:	Barbados
Capital:	Bridgetown
Terrain:	Relatively flat; rises gently to central highland region
Area:	166 mi² (430 km²)
Climate:	Tropical; rainy season (June to October)
Main river:	Constitution
Highest elevation:	Mount Hillaby, 1,115 ft (340 m)
Lowest elevation:	Atlantic Ocean, sea level
Form of government:	Constitutional monarchy
Head of state:	British monarch, represented by the governor general
Head of government:	Prime minister
Administrative areas:	11 parishes
Legislature:	Parliament consisting of the Senate with 21 members and the House of Assembly with 30 members serving five-year terms
Court system:	Supreme Court of Judicature
Armed forces:	610 troops
National holiday:	Independence Day - November 30 (1966)
Estimated 2010 population:	284,000
Population density:	1,711 persons per mi² (660 per km²)
Population distribution:	60% rural, 40% urban
Life expectancy in years:	Male, 72; female, 78
Doctors per 1,000 people:	1.2
Birth rate per 1,000:	13
Death rate per 1,000:	8
Infant mortality:	13 deaths per 1,000 live births
Age structure:	0-14: 20%; 15-64: 69%; 65 and over: 11%
Internet users per 100 people:	74
Internet code:	.bb
Language spoken:	English
Religions:	Protestant 63.4%, Roman Catholic 4.2%, other Christian 7%, other 25.4%
Currency:	Barbados dollar
Gross domestic product (GDP) in 2008:	$3.62 billion U.S.
Real annual growth rate (2008):	1.5%
GDP per capita (2008):	$12,846 U.S.
Goods exported:	Chemicals, electrical parts, molasses, rum, sugar
Goods imported:	Chemicals, food, machinery, transportation equipment
Trading partners:	Canada, Japan, Trinidad and Tobago, United Kingdom, United States

A Barbadian farmer inspects his sugar-cane crop as Caribbean waves break on a reef in the background. More than half the nation's farmland is used to grow sugar. Processing plants on the island produce refined sugar, molasses, and rum from the sugar cane.

Barbados lies about 250 miles (400 kilometers) northeast of Venezuela. It is the easternmost island in the West Indies. Profits from tourism and foreign investment have helped develop Barbados's economy.

the federation broke up in 1962, when Jamaica and Trinidad and Tobago became independent.

Barbados itself became an independent nation in 1966. Over the next 10 years, the Democratic Labor Party (DLP) held the majority of the seats in the House of Assembly, but in 1976, the Labor Party won a majority of seats. Control of the government has passed between the two parties several times since then.

People and economy

Barbados is one of the most densely populated islands in the world. About 80 percent of the people are descended from slaves brought to the island from Africa between 1636 and 1833. In 1833, slavery was abolished throughout the British Empire. More than 15 percent of the people are of mixed African and British ancestry. About 4 percent are of European—chiefly British—descent.

Barbados's economy is based on manufacturing and processing, tourism, and agriculture. Sugar cane, introduced to the island about 1640, is the country's chief agricultural product. In May, at the peak of the sugar-harvesting season, most of the island's farmers work on the sugar plantations. The islanders also raise avocados, carrots, cucumbers, okra, string beans, tomatoes, and yams.

BELARUS

Belarus *(behl uh ROOS)*—formerly Byelorussia—is an independent country and a member of the Commonwealth of Independent States (CIS), which was formed in late 1991. The country was formerly a republic of the now-defunct Soviet Union.

The dense hardwood forests that cover much of Belarus make lumbering an important industry. Timber, used in the manufacture of furniture, is floated to market down the Dnieper, Western Dvina, and Nyoman rivers.

The wet summers and sandy soils of Belarus are ideal for growing potatoes for use as animal feed as well as for human consumption. Other crops include barley, rye, sugar beets, and wheat.

The country is famous for the heavy-duty trucks and tractors it produces. Belarus also manufactures food products, metal-cutting tools, and such consumer goods as bicycles, motorcycles, refrigerators, and television sets.

The land that is now Belarus was first settled by East Slavic tribes in the 400's. In the 800's, the region came under the influence of the Kievan state, and during the 1100's, it was subdivided into several Kievan principalities. In the early 1300's, Belarus became part of the grand duchy of Lithuania, which merged with Poland in 1569.

Following the union of Lithuania and Poland, Belarus participated in the Polish Renaissance culture of the 1500's and 1600's. During this period, about 25 percent of its people became Roman Catholics. In the late 1700's, when Austria, Prussia, and Russia divided Poland up among themselves, Belarus became part of the Russian Empire.

During World War I (1914-1918) and the Soviet-Polish War (1919-1920), Belarus suffered great devastation. Communists formed the Byelorussian Republic in 1919, and in 1922, it joined the Union of Soviet Socialist Republics. The Treaty of Riga had awarded the western part of the republic to Poland in 1921, but the area was returned to the Soviets in 1939.

FACTS

Official name:	Respublika Byelarus (Republic of Belarus)
Capital:	Minsk
Terrain:	Generally flat and contains much marshland
Area:	80,155 mi² (207,600 km²)
Climate:	Cold winters, cool and moist summers; transitional between continental and maritime
Main rivers:	Dnieper, Neman (Nyoman), Western Dvina
Highest elevation:	Dzerzhinskaya Gora, 1,135 ft (346 m)
Lowest elevation:	Neman River, 295 ft (90m)
Form of government:	Republic
Head of state:	President
Head of government:	Prime minister
Administrative areas:	6 voblastsi (provinces), 1 horad (municipality)
Legislature:	Natsionalnoye Sobranie (National Assembly) consisting of the Soviet Respubliki (Council of the Republic) with 64 members serving four-year terms and the Palata Pretsaviteley (Chamber of Representatives) with 110 members serving four-year terms
Court system:	Constitutional Court, Supreme Court
Armed forces:	72,900 troops
National holiday:	Independence Day - July 3 (1944)
Estimated 2010 population:	9,577,000
Population density:	119 persons per mi² (46 per km²)
Population distribution:	73% urban, 27% rural
Life expectancy in years:	Male, 64; female, 76
Doctors per 1,000 people:	4.8
Birth rate per 1,000:	10
Death rate per 1,000:	14
Infant mortality:	6 deaths per 1,000 live births
Age structure:	0-14: 15%; 15-64: 70%; 65 and over: 15%
Internet users per 100 people:	30
Internet code:	.by
Languages spoken:	Belarusian, Russian
Religions:	Eastern Orthodox 80%, Roman Catholic 14%, Protestant 2%, other 4%
Currency:	Belarusian ruble
Gross domestic product (GDP) in 2008:	$60.29 billion U.S.
Real annual growth rate (2008):	9.2%
GDP per capita (2008):	$6,270 U.S.
Goods exported:	Chemicals, machinery, petroleum products, vehicles (especially tractors and trucks)
Goods imported:	Iron and steel, machinery, petroleum products, transportation equipment
Trading partners:	Germany, Netherlands, Poland, Russia, Ukraine

Belarus, also known as White Russia, is a land of hills and marshes. It is surrounded by Poland in the west, Lithuania and Latvia in the north, Russia in the east, and Ukraine in the south.

Bison roam in the ancient Belovezha Forest, a nature preserve in the Brest region of Belarus.

Because it lay along the direct route between Berlin and Moscow, Belarus suffered great losses during World War II (1939-1945) as the German Army advanced along the Western Front. During the Nazi occupation, 1.5 million Belarusians were permanently moved east of the Volga River, and a large segment of the Jewish population fled abroad. More than three-quarters of the republic's towns and cities were destroyed.

After World War II, Belarus rebuilt and restored its urban areas. By the late 1970's, it was one of the leading Soviet republics in urban and industrial growth.

An explosion at the Chernobyl nuclear power plant in Ukraine in 1986 had a major impact on Belarus. The winds caused about 70 percent of the radioactive fallout to land on Belarus. The radiation contaminated the republic's food and water supplies and caused many health problems.

Belarus was under the control of the Soviet central government until the early 1990's. In 1991, Belarus declared its independence. In December, it was one of three former Soviet republics, with Russia and Ukraine, to establish the Commonwealth of Independent States. In 1994, Belarus announced an agreement to merge its economy with Russia's. Many Belarusians opposed the union, and it was postponed. The parliament outlawed use of the Russian ruble for domestic transactions.

Aleksandr Lukashenko became president of Belarus in 1994, the same year the country adopted a new constitution. In 1996, Lukashenko pushed through a referendum for a new constitution that would expand his power. The referendum passed, but opponents of Lukashenko said the vote had been falsified. Lukashenko was reelected in 2001, 2006, and 2010. Observers claimed that all of these elections were marred by irregularities.

BELGIUM

Belgium, a small country in northwestern Europe, borders France, the Netherlands, and Germany. A narrow strip of the North Sea separates Belgium from the United Kingdom. Its geographical position helped Belgium become an important European industrial and trade center, but its location has also made it a battleground for warring nations. Belgium suffered great destruction during World War I (1914-1918) and World War II (1939-1945).

With about 893 persons per square mile (345 per square kilometer), Belgium is one of the most densely populated countries in the world. Most of the people are divided into two major groups, the Flemings and the Walloons. The Flemings live in northern Belgium—Flanders—and speak Dutch; the Walloons live in southern Belgium—Wallonia—and speak French. Both Flemings and Walloons live in Brussels, the nation's capital. Dutch and French are both official languages of the country, but the differences in language and other cultural traditions have long caused friction between the two groups.

Despite its small size, Belgium has a varied landscape. Dunes and beaches line its northern coast, while forest-covered hills extend across much of the southeastern part of the country. The central region, which has Belgium's best soil, is also the site of many of the nation's largest cities, including Brussels and Liège. Several large rivers serve as transportation routes. The country has no large natural lakes, but engineers have created several lakes in the south by damming the rivers.

Belgium has a rich architectural and artistic heritage. Stately buildings and churches erected hundreds of years ago still stand in many towns and cities. Museums are filled with works by such outstanding Flemish artists as Jan van Eyck, Pieter Bruegel the Elder, and Peter Paul Rubens. In literature, too, Belgium has made contributions—both in French and Dutch. The Flemish writer Maurice Maeterlinck won the Nobel Prize for literature in 1911 for his plays, including *The Blue Bird*, written in French. The Flemish poet and novelist Hugo Claus is generally considered the leading Belgian writer in Dutch since the mid-1900's.

Since World War II, Belgium has played a leading role in European economics and politics. It is a member of Benelux—the economic union of Belgium, the Netherlands, and Luxembourg—and is the headquarters for the European Union and the North Atlantic Treaty Organization (NATO). The futuristic sculpture called *Atomium,* symbol of the 1958 Belgian World's Fair, represented Belgium's new position in European life.

BELGIUM TODAY

Until it gained its independence from the Netherlands in 1830, Belgium was governed by many different foreign rulers, including the Romans, the Franks, the Spanish, the Austrians, the French, and the Dutch. The Belgians revolted against Dutch rule in August 1830 and declared their independence on October 4. In 1831, Belgium adopted a constitution and chose Prince Leopold of Saxe-Coburg as its king. Leopold was the uncle of Queen Victoria of the United Kingdom.

Belgium is a constitutional monarchy, but the king has little real power. Executive power lies in the hands of the prime minister and the members of a cabinet called the Council of Ministers, which consists of equal numbers of Dutch-speaking and French-speaking members. The prime minister holds office as long as the support of Belgium's two-house parliament lasts.

Political and cultural regions
On the local level, Belgium is divided into 10 provinces, each of which is headed by a governor and a council. Belgium is also divided into three cultural communities—those of the Flemings, Germans, and Walloons—and three economic regions—Flanders, Wallonia, and Brussels.

Efforts by the Flemings and Walloons to preserve their separate cultural identities have sometimes bordered on civil war. Their conflict not only divides the country culturally, but politically and economically as well. In an effort to seek a solution to the division between the communities, the Belgian government granted both groups limited self-rule in 1980.

FACTS

Official name:	Royaume de Belgique/Koninkrijk Belgie (Kingdom of Belgium)
Capital:	Brussels
Terrain:	Flat coastal plains in northwest, central rolling hills, rugged mountains of Ardennes Forest in southeast
Area:	11,787 mi² (30,528 km²)
Climate:	Temperate; mild winters, cool summers; rainy, humid, cloudy
Main rivers:	Schelde, Sambre, Meuse
Highest elevation:	Botrange Mountain, 2,277 ft (694 m)
Lowest elevation:	North Sea, sea level
Form of government:	Parliamentary democracy
Head of state:	Monarch
Head of government:	Prime minister
Administrative areas:	10 provinces
Legislature:	Parliament consisting of a Senaat or Senat, (Dutch and French terms for Senate) with 71 members serving four-year terms and a Kamer van Volksvertegenwoordigers or Chambre des Representants (Dutch and French terms for Chamber of Deputies) with 150 members serving four-year terms
Court system:	Hof van Cassatie, in Dutch, or Cour de Cassation, in French (both meaning Supreme Court)
Armed forces:	38,800 troops
National holiday:	Ascension to the throne of King Leopold I - July 21 (1831)
Estimated 2010 population:	10,520,000
Population density:	893 persons per mi² (345 per km²)
Population distribution:	97% urban, 3% rural
Life expectancy in years:	Male, 77; female, 82
Doctors per 1,000 people:	4.2
Birth rate per 1,000:	11
Death rate per 1,000:	10
Infant mortality:	4 deaths per 1,000 live births
Age structure:	0-14: 17%; 15-64: 66%; 65 and over: 17%
Internet users per 100 people:	67
Internet code:	.be
Languages spoken:	Dutch, French, German (all official)
Religions:	Roman Catholic 75%, other (Protestant, Jewish, Muslim, Anglican, Greek, Russian Orthodox) 25%
Currency:	Euro
Gross domestic product (GDP) in 2008:	$499.79 billion U.S.
Real annual growth rate (2008):	1.3%
GDP per capita (2008):	$47,645 U.S.
Goods exported:	Chemicals, diamonds, machinery, metals and metal products, transportation equipment
Goods imported:	Chemicals, diamonds, machinery, petroleum products, transportation equipment
Trading partners:	France, Germany, Netherlands

BELGIUM

North Sea

NETHERLANDS

FRANCE

GERMANY

LUXEMBOURG

FRANCE

The imposing Palais de Justice, or law courts building, is one of Brussels's most important landmarks. The broad Rue de la Régence, in the foreground, runs northeast toward the Palais de Roi, the residence of Belgium's monarch.

Belgium is a small but densely populated country in northwestern Europe. It borders France, Germany, Luxembourg, and the Netherlands. Belgium's moist, moderate climate benefits the country's farmland.

Land regions

Belgium's landscape varies greatly for so small a country. Belgium's four main land regions are the Coastal and Interior Lowlands, the Kempenland, the Central Low Plateaus, and the Ardennes.

The Coastal and Interior Lowlands extend across most of northern Belgium. Sandy beaches line the North Sea coast, while sea walls and drainage canals protect the nearby lowlands from flooding. The fertile inland region provides good farmland.

The Kempenland, also called the *Campine,* is a mining and industrial area in northeastern Belgium. Many of the birch forests that grew there until the early 1900's have been cleared and replanted with fast-growing evergreens for timber harvest.

The Central Low Plateaus in central Belgium have the country's richest soil. This region is also the site of many of the nation's largest cities.

The Ardennes region, which covers southeastern Belgium, consists mainly of forested hills separated by winding rivers. Many deer and wild boars roam these forests.

FLEMINGS AND WALLOONS

A visitor traveling through Belgium could hardly fail to notice that people in the northern part of the country speak Dutch, while those in the south speak French. The Dutch-speaking Flemings, who live in the northern region of Flanders, make up almost 60 percent of the Belgian population. The French-speaking Walloons, who live in the southern region of Wallonia, make up about 30 percent of the population. Although the capital city of Brussels is located in Flanders, both languages are spoken there. The language difference has been a source of political and economic conflict between the two regions for more than 175 years.

Roots of the conflict

The conflict began in 1830, when Belgium achieved independence and French was recognized as the new nation's only official language. The Flemings were poorly represented in the Belgian government, and many Dutch speakers protested the domination of Belgium by French speakers. The Walloons also dominated the nation economically. Between 1830 and 1870, heavy industry developed rapidly in Wallonia, and it became one of Belgium's major economic centers.

In the late 1800's, the Flemings won recognition of Dutch as the nation's second official language. However, conflict between the Flemings and the Walloons continued into the 1900's as each group sought to advance its own economic and cultural interests.

Tensions between the two groups increased during the 1960's. The Flemings held mass demonstrations to demand political equality and cultural independence. In the 1970's, energy crises caused a decline in industrial Wallonia's wealth, and high unemployment followed. As a result, the economic and political balance shifted to the Flemings.

Signs in Dutch and French in Brussels indicate the nation's bilingual status.

Mass demonstrations staged over many years reflect the continuing tensions between the Flemings and the Walloons.

The situation today

In 1971, Parliament revised the Constitution to divide the nation into three cultural communities: Dutch speakers, French speakers, and German speakers. Each community was given its own cultural council, which must approve all legislation dealing with language, education, or other cultural matters. The revised Constitution also set up three economic regions in Belgium: Flanders, Wallonia, and Brussels. Finally, in 1980, the Belgian government granted limited self-rule to Flanders and Wallonia.

Today, the Belgian government and most businesses use both French and Dutch. In Flanders, schools teach in Dutch. Language-based political parties have had an important influence on Belgian politics. After inconclusive elections in June 2010, the country's Dutch-speaking and French-speaking political parties were not able to form a government. Belgium set a modern-day world record of 535 days without an official government. The parties reached an agreement in November 2011 that six parties would form a coalition government.

Antwerp's Grote Markt flourishes in the new-found prosperity of Flanders. During the 1500's, Antwerp's harbor helped the city become one of the richest trading centers in the world. Today, the harbor at Antwerp is still among the world's largest, and the city also benefits from the petrochemical plants established in Flanders.

Major urban areas
- ● More than 150,000 inhabitants
- ● 100,000 to 150,000 inhabitants
- ○ Less than 100,000 inhabitants

- Dutch (Flemish) speaking areas
- French speaking areas
- Dutch and French speaking areas
- German speaking areas

In Belgium, one of the most densely populated countries of the world, more than 95 percent of the people live in cities and towns. A jagged east-west line across the country's map indicates the division between French-speaking Wallonia in the south and Dutch-speaking Flanders in the north. The small German-speaking community in the east maintains its own cultural identity.

The Mons region of French-speaking Wallonia reflects the decay of the declining industrial area. Since the 1960's, Wallonia has lost its commercial dominance over Flanders.

MEDIEVAL CITIES OF FLANDERS

Until 1830, Flanders was a political unit that included areas that are now part of France and the Netherlands, as well as the northern half of Belgium. During the Middle Ages, Flanders was one of Europe's most important economic centers, largely as a result of its cloth-making industry. The area's position at the crossroads of trading routes to the south and east also contributed to its development. The wealth of the merchant class that settled in Ghent, Bruges, Antwerp, and Ostend is still reflected in these cities.

Ghent

Ghent, which lies at the fork of the Schelde and Leie rivers, has been the capital of Flanders since the 1100's. By the 1400's, the city was one of the most important trading towns in the Hanseatic League and the center of the Flanders cloth trade.

For many years afterward, however, Ghent's economy was destroyed by revolutions and war. The cloth industry eventually revitalized the city's economy in the 1800's. Ghent is connected to the North Sea by a ship canal built in 1886, and the city now ranks as a leading seaport.

Ghent's historic buildings, particularly the guildhalls that line its squares, testify to its former position as a great commercial city. Ghent was also a renowned art center, and numerous paintings, including many by famous Flemish artists, hang in its churches and museums.

The city's medieval buildings, along with some 200 bridges spanning its waterways, make a timeless setting. The fortified castle of Gravensteen, which overlooks Ghent, presents an excellent view of the city's old section.

The view of Bruges from above shows why many people consider it to be the most beautiful city in Belgium. Filled with historic buildings and threaded by picturesque canals, Bruges has preserved much of its medieval past.

Ostend is a popular seaside resort as well as one of Belgium's major seaports. A fortified town in the 1500's, Ostend became a fashionable resort in the late 1800's.

Ghent is an important Belgian seaport. The city lies at the fork of the Schelde and Leie rivers. Many gabled houses along the waterfront date from the 1400's and 1500's.

Bruges

From 1240 to 1426, Bruges was one of the most important cities in Europe. As a member of the Hanseatic League, Bruges rivaled the great trading empire of Venice. Ships carrying cloth, silks, gold, salt, and spices sailed directly into the port at Bruges via an inlet that connected the city with the North Sea. Much of the wealth of the city, however, came from the wool trade, which Bruges essentially monopolized.

The city's importance decreased as the Hanseatic League declined and the inlet dried up. Bruges finally experienced an economic revival in the 1900's, due to the construction of the Bruges maritime canal and the expansion of the port of Zeebrugge.

Today, although Bruges is an important lacemaking center, tourism has replaced trade as the city's major economic activity. Thousands of visitors come to Bruges to see the magnificent buildings dating back to the Middle Ages and their beautiful carvings and paintings. Such buildings include the Market Hall, built in the 1200's, and the city's Gothic town hall, built in the 1300's. The many bridges that cross the network of canals gave the city its name—*Bruges* means *bridges*.

Waterways link the historic lacemaking town of Mechelen with other communities of northern Belgium.

BRUSSELS

Brussels, Belgium's capital, ranks as the nation's fifth largest city, but Brussels and its suburbs make up the country's largest metropolitan area. While both Dutch and French are used for education and public communication in metropolitan Brussels, French is the everyday language of most of the people. The city is called *Brussel* in Dutch and *Bruxelles* in French.

History of the city

Historians do not know when Brussels was founded. By the A.D. 900's, the city had become an important stopping point on trade routes linking western Germany and northern France. By the 1200's, Brussels itself was an important center of trade and industry, famous for its fine tapestries and other textile products.

For centuries thereafter, Brussels was part of empires controlled by foreign rulers, including the Burgundians, Spaniards, Austrians, French, and Dutch. The city became the capital of Belgium when the country gained independence in 1830. German troops occupied Brussels during World War I (1914–1918) and again during World War II (1939–1945), but the city suffered little physical damage in these wars.

Sometimes called the "capital of Europe," Brussels today is a center of international economic and political activity. The European Union (EU) and the North Atlantic Treaty Organization (NATO) are based in or near the city. Many of Brussels's residents work for the government or for EU agencies.

With its large office buildings, cafes, hotels, shops, and network of wide expressways, Brussels appears to be a very modern city. However, many reminders of its past are found in the city's old section.

Beautiful lace has been one of the most famous products of Belgium since the 1500's, when Italy and Belgium were the chief centers of early lacemaking. A plaque at the left marks the residence of the famous French author Victor Hugo, who lived in Brussels during his exile from France.

The Atomium was built for Expo 58, the world's fair held in Brussels in 1958. The monument is a giant model of a unit cell of an iron crystal. Atomium has become a symbol of the city and one of its most popular tourist attractions.

The lower city

The oldest section of the city, called the *lower city,* lies in the center of Brussels. At its heart stands the Grand' Place, the main square of Brussels. This marketplace is bordered by elaborately decorated buildings constructed during the late 1600's to house merchant and craft guilds.

These buildings were erected after French cannon fire destroyed most of the square in 1695. Only Brussels's town hall, the Hôtel de Ville—dating from the 1400's—survived the destruction. Atop the building's graceful spire, a statue of Saint Michael, the city's patron saint, faces the square.

Brussels's best-loved landmark, the *Manneken-Pis,* can be found on a street near the Grand' Place. Known as the "Oldest Citizen of Brussels," this bronze statue of a small, naked boy symbolizes the city's spirit of cheerful independence.

The upper city

The *upper city*, which lies east of the lower city, contains many important buildings erected during the 1800's and early 1900's, including the royal palace and the Parliament building, as well as elegant residential neighborhoods. Near the royal palace stands the Palais de Justice, which houses Belgium's highest court. This massive building can be seen from many parts of Brussels.

The Cathedral of Saint Michael, north of the Palais de Justice, is renowned for its stained-glass windows, dating from the early 1500's. The church, which stands on the site of an earlier chapel, was itself begun in 1226.

Art and education

Many people come to Brussels to see the magnificent art collections in its museums. The Museum of Fine Arts features works by Belgian artists of the 1400's through 1800's, while the Museum of Modern Art houses more recent works. Other cultural attractions include the Museum of Natural History, the Albert I Library, and the Théâtre de la Monnaie, which offers performances of operas and ballets.

The Free University of Brussels is actually two institutions. Consistent with the city's official bilingual status, the university has one division for French-speaking students and another for those who speak Dutch.

The Brussels town hall stands on one side of the Grand' Place, one of the most beautiful and spacious town squares in the world. The town hall was completed in the mid-1400's.

The Cité Berlaymont is the headquarters of the European Commission, which administers the European Union (EU). Many other international agencies, including NATO and Benelux, also meet in Brussels.

ECONOMY

Belgium has a highly developed economy based on *free enterprise,* a system in which businesses operate with little government control. But the government owns and manages parts of the transportation and communication systems. The government also provides basic social services and medical insurance coverage for all citizens.

Belgium's economy has undergone considerable restructuring. Traditional industries, such as coal mining, have declined. In addition, important industries that were once concentrated in Wallonia, such as the production of steel, have moved to Flanders. Today, service industries employ most of Belgium's workers. Important service industries include such areas as education, health care, and government. In addition, finance, restaurants, and wholesale and retail trade are important. Belgium's cities are the centers of most of its service industries.

Economy in Wallonia and Flanders

The province of Hainaut in Wallonia is the country's oldest industrial region. Coal mines along the Sambre River once helped make southern Belgium a prosperous area, but since 1958, many mines have closed down due to high production costs and the exhaustion of the coal deposits.

Steel industries located in Wallonia declined during the worldwide steel crisis of the 1970's and were not modernized thereafter. These plants used coal rather than more-efficient petroleum.

At the same time, much of the farmland in Flanders was transformed into modern industrial areas. The Belgian steel industry, which had been concentrated near the coal mines in Wallonia, moved to Flanders. The industry now uses imported petroleum—instead of coal—to power the steel-making process, so the new steel plants are located near the North Sea. The ports of Antwerp, Ghent, and Zeebrugge have all benefited from the expansion of industry in Flanders.

A carpet factory in Mouscron, western Belgium, produces one of the country's chief textile products. Belgium's long-established textile industry also exports synthetic fibers, lace, and linen.

In the 1970's, the Belgian government established Flanders and Wallonia as relatively independent economic regions. However, Belgium's future economic development depends on the government's success in achieving economic equality for the inhabitants of both areas. Improved cooperation between the Flemings and Walloons is also necessary to reduce Belgium's high rate of unemployment, keep inflation down, and further stimulate the economy's growth.

Industry and agriculture

Steel production is the country's most important manufacturing industry. Belgium is among the world's leading nations in *per capita* (per person) steel production.

The chemical industry manufactures basic chemicals as well as pharmaceutical products and plastics. Belgium's textile products include its world-famous lace and linen. Food processing, including the manufacture of Belgian chocolates, also ranks as an important industry. There are also many breweries in Belgium, including some that are located in monasteries.

Antwerp is Belgium's main seaport and the economic center of Flanders. In addition to its vast trade, the city has many industries, including sugar refining, lacemaking, brewing, and shipbuilding. Antwerp is also a major diamond center.

Belgium depends heavily on foreign trade because the country has very few natural resources. The nation's main trading partners are other members of the European Union, especially Germany, the Netherlands, and France.

Machines and other engineering goods make up the largest share of Belgium's foreign trade. The country also imports chemicals, diamonds, grains, and petroleum. Other major exports include chemicals, diamonds, glass products, processed foods, steel, and textiles.

Farmers make up less than 2 percent of the Belgian work force, but they produce most of the country's food needs. Dairy farming and livestock production account for more than two-thirds of Belgium's farm income. The nation's agricultural land is used to grow barley, potatoes, sugar beets, and wheat. Belgium also produces large quantities of flowers, especially azaleas, as well as fruits and vegetables.

Barges crowd Antwerp's busy port which ranks as one of Europe's largest. Located near the mouth of the Schelde River, Antwerp has a varied industry, which includes brewing, diamond cutting and trading, distilling, shipbuilding, and sugar refining.

Highly productive Belgian farms average about 58 acres (23 hectares) in size. Most farms are run by families, many of whom rent the land. Modern farm machinery has replaced the horses which used to pull plows and other farm implements until the 1960's.

BELIZE

Belize *(buh LEEZ)* is a tiny country in Central America. It lies close to Honduras on the southeast coast of the Yucatán Peninsula, south of Mexico and east of Guatemala.

Until 1973, Belize was called British Honduras. The United Kingdom ruled the country from the mid-1800's until 1981, when Belize became an independent nation. However, British troops remained until 1993 because Guatemala claimed Belize. Guatemala gave up this claim in 1991.

As a former British colony, Belize is now part of the Commonwealth of Nations. It is a constitutional monarchy with the British monarch as its head of state. The government is a parliamentary democracy.

Belize's legislature consists of a House of Representatives and a Senate. The people elect the members of the House, and the leader of the majority party becomes the prime minister. With the help of a cabinet, the prime minister runs the government. A governor general represents the British monarch.

The land and economy

Most of coastal Belize is warm, swampy lowland. Inland, the north is flat, but the land in the south rises to the low peaks of the Maya Mountains. The country is covered by many forests, though large areas have been cut for lumber. Pine and such hardwoods as mahogany and cedrela are important forest products.

Belize is a developing country with an economy based largely on agriculture. The country's primary crops include bananas, grapefruit, oranges, and sugar. Belize's forests contribute a small amount to the country's economy. In 2006, Belize began producing oil.

Most industries in Belize are small. These industries process food, produce cement and clothing, and refine sugar.

Belize's location on the Caribbean Sea helps contribute to its economy. The sea provides fish and shrimp. Tourism is becoming more important to Belize's economy. Most tourists come from the United States.

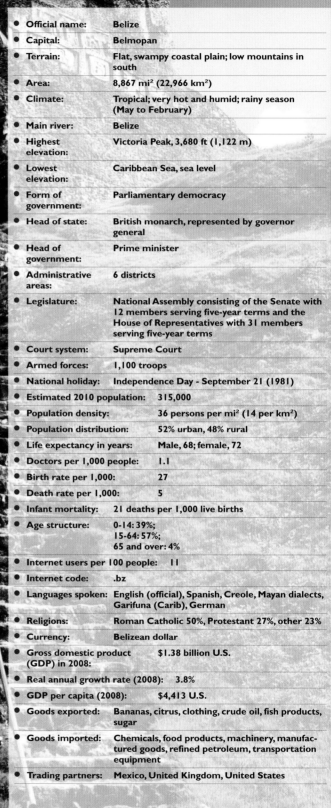

FACTS

Official name:	Belize
Capital:	Belmopan
Terrain:	Flat, swampy coastal plain; low mountains in south
Area:	8,867 mi² (22,966 km²)
Climate:	Tropical; very hot and humid; rainy season (May to February)
Main river:	Belize
Highest elevation:	Victoria Peak, 3,680 ft (1,122 m)
Lowest elevation:	Caribbean Sea, sea level
Form of government:	Parliamentary democracy
Head of state:	British monarch, represented by governor general
Head of government:	Prime minister
Administrative areas:	6 districts
Legislature:	National Assembly consisting of the Senate with 12 members serving five-year terms and the House of Representatives with 31 members serving five-year terms
Court system:	Supreme Court
Armed forces:	1,100 troops
National holiday:	Independence Day - September 21 (1981)
Estimated 2010 population:	315,000
Population density:	36 persons per mi² (14 per km²)
Population distribution:	52% urban, 48% rural
Life expectancy in years:	Male, 68; female, 72
Doctors per 1,000 people:	1.1
Birth rate per 1,000:	27
Death rate per 1,000:	5
Infant mortality:	21 deaths per 1,000 live births
Age structure:	0-14: 39%; 15-64: 57%; 65 and over: 4%
Internet users per 100 people:	11
Internet code:	.bz
Languages spoken:	English (official), Spanish, Creole, Mayan dialects, Garifuna (Carib), German
Religions:	Roman Catholic 50%, Protestant 27%, other 23%
Currency:	Belizean dollar
Gross domestic product (GDP) in 2008:	$1.38 billion U.S.
Real annual growth rate (2008):	3.8%
GDP per capita (2008):	$4,413 U.S.
Goods exported:	Bananas, citrus, clothing, crude oil, fish products, sugar
Goods imported:	Chemicals, food products, machinery, manufactured goods, refined petroleum, transportation equipment
Trading partners:	Mexico, United Kingdom, United States

The people

The ancient Maya spread into what is now Belize from the northern part of the Yucatán Peninsula. Some Belizeans still speak Mayan Indian languages today. However, English is the country's official and most commonly used language, followed by Spanish. Many people speak both English and Spanish.

Belize is an ethnically diverse country. About half of the population are *mestizos* (people of mixed European and American Indian ancestry). About a fourth of the people are Creoles (people of mixed African and European ancestry). Most of the rest are Maya Indians or Garifuna (people of mixed African and American Indian ancestry).

Most Belizeans are poor. About half live in urban areas, and about half live in rural areas. Unemployment is high in the cities, and farm production is low in the rural areas.

Belize is one of the smallest countries in Central America and the most thinly populated. More than half the people live along the Caribbean coast. Offshore lies one of the longest barrier coral reefs in the world. The Belize River cuts across the country.

Mayan ruins are preserved in Caracol in Belize. The Temple of the Wooden Lintel was one of the earliest Mayan structures, dating from the A.D. 500's or 600's.

BENIN

The horned headdress that adorns this young woman may reflect her spiritual beliefs. Many of Benin's people practice animism, the belief that all things in nature have spirits.

The long, narrow western African country of Benin (beh NEEN), known as Dahomey until 1975, stretches 415 miles (668 kilometers) inland from the Gulf of Guinea. More than 400 years ago, Europeans established slave-trading posts along the coast, and the kingdom of Dahomey controlled the region with power based largely on the slave trade.

During the 1800's, the palm oil trade replaced the slave trade, and in 1851 France signed a trade agreement with the king of Dahomey. However, when Dahomey soldiers attacked French trading posts in 1892, France took over the area and, in 1904, made it a French territory. The French built roads and railroads that made Benin a crossroads for coastal road traffic and provided an outlet to the sea for inland territories. Self-government was granted to Dahomey in 1958, and in 1960 it became a fully independent nation.

The government has undergone frequent changes since independence, however. In 1972, a military group overthrew a civilian government, and a military government led by army officer Mathieu Kerekou took control of the nation's most important businesses. Three years later, the government changed the nation's name from Dahomey to Benin.

In 1990, Kerekou's government was dissolved, all political parties were legalized, and a temporary government was set up. Kerekou remained president, but Nicephore Soglo became prime minister and leader of Benin. The temporary government served until early 1991, when Soglo was elected president and a new legislature was chosen. Benin's new government ended nearly all governmental control of businesses. Kerekou was elected president in 1996 and reelected in 2001. Yayi Boni, a former banker, was elected to succeed Kerekou in 2006. He was reelected in 2011.

FACTS

Official name:	Republique du Benin (Republic of Benin)
Capital:	Porto-Novo (official), Cotonou (seat of government)
Terrain:	Mostly flat to undulating plain; some hills and low mountains
Area:	43,484 mi² (112,622 km²)
Climate:	Tropical; hot, humid in south; semiarid in north
Main rivers:	Ouémé, Okpara
Highest elevation:	Atacora Mountains, 2,083 ft (635 m)
Lowest elevation:	Atlantic Ocean, sea level
Form of government:	Republic
Head of state:	President
Head of government:	President
Administrative areas:	12 departments
Legislature:	Assemblee Nationale (National Assembly) with 83 members serving four-year terms
Court system:	Cour Constitutionnelle (Constitutional Court), Cour Supreme (Supreme Court), High Court of Justice
Armed forces:	4,800 troops
National holiday:	National Day - August 1 (1960)
Estimated 2010 population:	9,056,000
Population density:	208 persons per mi² (80 per km²)
Population distribution:	59% rural, 41% urban
Life expectancy in years:	Male, 56; female, 59
Doctors per 1,000 people:	Less than 0.05
Birth rate per 1,000:	40
Death rate per 1,000:	11
Infant mortality:	78 deaths per 1,000 live births
Age structure:	0-14: 44%; 15-64: 53%; 65 and over: 3%
Internet users per 100 people:	2
Internet code:	.bj
Languages spoken:	French (official), Fon and Yoruba (in south), tribal languages (at least 6 major ones in north)
Religions:	Christian 42.8%, Muslim 24.4%, Vodoun 17.3%, other 15.5%
Currency:	Communaute Financiere Africaine franc
Gross domestic product (GDP) in 2008:	$6.94 billion U.S.
Real annual growth rate (2008):	4.8%
GDP per capita (2008):	$860 U.S.
Goods exported:	Cotton, palm products, textiles
Goods imported:	Capital goods, food, petroleum products
Trading partners:	China, France

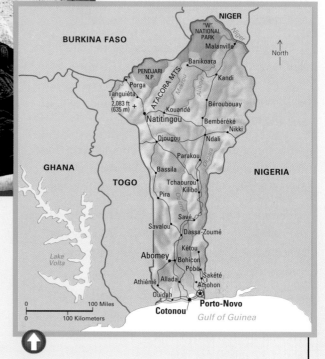

The production of wool is a traditional livelihood in Benin. A Bariba man of northern Benin lifts a skein of wool yarn from a dyeing vat.

Benin is a long, narrow country in western Africa. It extends 415 miles (668 kilometers) inland from the coast. Porto-Novo is the capital city, but most government activity occurs in Cotonou.

Benin is mainly an agricultural country. Cotton is the country's main export. Benin also exports products from its palm trees, such as palm oil and palm kernels. A few industrial plants reside in the south. Benin produces cement and other construction materials, food products, and textiles.

Most of the people of Benin work as farmers. Farmers in Benin grow beans, cashews, cassava, corn, peanuts, sorghum, and yams. They raise such livestock as cattle, chicken, goats, pigs, and sheep.

The people of Benin belong to some 60 ethnic groups. The largest group, which makes up about 60 percent of the population, consists of the closely related Fon and Adja. They live in southern Benin along with the Yoruba, who make up about 10 percent of the population. The Bariba, who also make up about 10 percent, are the largest group in the north.

Most people in Benin live in simple houses built by hand. However, some people, particularly in the cities, live in homes made of concrete.

The women of Benin often wear brightly colored dresses, and many of the men wear the *agbade*—an outfit of trousers, a full-length robe, and a short jacket. However, many people, particularly in southern Benin, wear clothing that is similar to that worn in the United States and Canada.

BERMUDA

It is not surprising that Bermuda is a popular resort area. This string of coral islands and islets is blessed with a mild climate, beautiful beaches, swaying palm trees, and colorful flowers. Every year, numerous tourists and honeymooners visit Bermuda.

Bermuda actually consists of about 140 islands and tiny islets, about 670 miles (1,080 kilometers) southeast of New York City in the North Atlantic Ocean. The islands form one of the most northerly coral island groups in the world, lying midway between Nova Scotia and the West Indies.

Only about 20 of the islands are inhabited. The four largest islands—Bermuda, St. George's, St. David's, and Somerset—extend in a chain about 22 miles (35 kilometers) long. These islands and several smaller ones are linked by bridges.

Bermuda's total land area is about 20 square miles (53 square kilometers). The island of Bermuda takes up about two-thirds of that area.

Bermuda has hills and ridges that rise up out of the sea as high as 260 feet (80 meters). Scenic beaches line the coasts, and caves are found throughout the islands.

Bermuda is noted for its dangerously narrow, winding roads. Only small automobiles are permitted, and the top speed limit is 35 kilometers (about 22 miles) per hour.

Bermuda's climate is mild, and rainfall is fairly plentiful. Rainfall is important because Bermuda has few underground sources of fresh water. Rain water is collected on rooftops and stored in tanks under buildings. To help keep the water supply pure, roofs are kept clean and whitewashed. Like the southeast coast of North America, Bermuda is sometimes struck by hurricanes in the fall.

About 65,000 people live on the 20 inhabited islands. People of African descent make up a majority of the population. People from Asia, Europe, and North America also live in Bermuda.

Fishing in Bermuda is chiefly a tourist sport, but a few Bermudians make their living by commercial fishing.

Bermuda is made up of about 140 islands and tiny islets; most are uninhabited. The capital city of Hamilton is on the island of Bermuda.

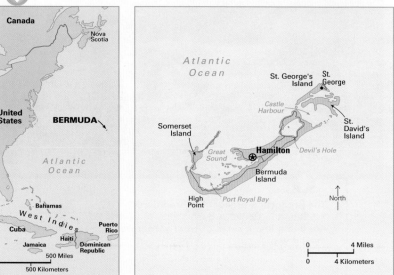

Tourism is the major source of income on the islands. Bermuda also attracts businesses from other countries by giving them tax breaks. For that reason, thousands of foreign companies operate in Bermuda, including many insurance and investment firms. There is little land that can be farmed, so Bermuda imports much of its food.

Government

Bermuda is a self-governing British dependency. The British monarch appoints the governor of Bermuda.

Bermuda's parliament, established in 1620, is the world's oldest British overseas parliament. It is made up of an 11-member appointed Senate and a 36-member elected House of Assembly.

Bermuda's government was largely controlled by the European minority for most of its history. During the late 1960's and the 1970's, many people of African descent protested against this control. In 1998, the Progressive Labour Party, mostly representative of Bermudians of African descent, won control of the government for the first time. The party remained in power following general elections in 2003 and 2007.

History

The islands were named for Juan de Bermúdez, a Spaniard who discovered them in the 1500's. In 1609, a ship carrying colonists to Virginia was destroyed in a violent storm, and the passengers took refuge in Bermuda. All except two sailed away the next year. Those two people became Bermuda's first permanent settlers.

In 1684, the British government took control of the islands. The English settlers kept African slaves as domestic servants and boat builders. From the 1680's to the early 1800's, Bermuda carried on a thriving trade with the West Indies and North America. The salvage of shipwrecks, blockade running, and smuggling are all part of Bermuda's history. It has also been the site of U.S. military bases. Today the North Atlantic Treaty Organization (NATO) provides for Bermuda's defense.

The beach resort on the island of St. George's caters to many of the tourists who visit Bermuda on giant luxury cruise ships each year.

BHUTAN

Bhutan (boo TAHN) is a small mountain kingdom that lies in the eastern Himalaya between India and Tibet. Bhutan is a constitutional monarchy. It had long been a *hereditary* (inherited) monarchy, but in 2008, a new constitution made the country's government more democratic.

Bhutia fathers often carry their young children on their backs. The dress of the Bhutias, who make up more than half the population of Bhutan, reveals a Tibetan influence.

Bhutan's early history is uncertain. The country's original settlers, the Bhutia Tephoo, were conquered by Tibetan invaders in the A.D. 800's. In the 1600's, a Tibetan *lama* (Buddhist monk) took control of the country's religious and political affairs. In 1907, Ugyen Wangchuk became Bhutan's first king.

Over the past 200 years, both India and the United Kingdom have taken control of Bhutan's affairs. During the 1700's and 1800's, the British controlled Bhutan's foreign relations. In 1910, the British Indian government took over but left Bhutan in control of its own internal affairs. In 1949, India agreed to help Bhutan develop its economy. Bhutan strengthened its ties to India in 1959 and began to modernize its economic, educational, and health care systems.

Landscape and economy

Bhutan has three major land regions. Along the Indian border lies a region of plains and river valleys with a hot, humid climate where bananas, citrus fruits, and rice are grown.

The mid-Himalayan region is covered with forests of ash, oak, poplar, and willow trees. The climate in this region is moderate.

In the northernmost region of Bhutan, the moun-

FACTS

Official name:	Kingdom of Bhutan
Capital:	Thimphu
Terrain:	Mostly mountainous with some fertile valleys and savanna
Area:	14,824 mi² (38,394 km²)
Climate:	Tropical in southern plains; cool winters and hot summers in central valleys; severe winters and cool summers in Himalaya
Main rivers:	Wong Chu, Sankosh, Tongsa, Bumtang, Manas, Kuru
Highest elevation:	Kula Kangri, 24,783 ft (7,554 m)
Lowest elevation:	150 ft (46 m) in the south
Form of government:	Constitutional monarchy
Head of state:	Monarch
Head of government:	Prime minister
Administrative areas:	20 dzongkhag (districts)
Legislature:	Parliament consisting of the National Council with 25 members serving four-year terms and the National Assembly with 47 members serving five-year terms.
Court system:	Supreme Court of Appeal, High Court
Armed forces:	N/A
National holiday:	National Day - December 17 (1907)
Estimated 2010 population:	696,000
Population density:	47 persons per mi² (18 per km²)
Population distribution:	66% rural, 34% urban
Life expectancy in years:	Male, 66; female, 67
Doctors per 1,000 people:	Less than 0.05
Birth rate per 1,000:	25
Death rate per 1,000:	7
Infant mortality:	45 deaths per 1,000 live births
Age structure:	0-14: 31%; 15-64: 63%; 65 and over: 6%
Internet users per 100 people:	6
Internet code:	.bt
Languages spoken:	Dzongkha (official), Tibetan dialects, Nepalese dialects
Religions:	Lamaistic Buddhist 75%, Indian- and Nepalese- influenced Hinduism 25%
Currency:	Ngultrum
Gross domestic product (GDP) in 2008:	$1.37 billion U.S.
Real annual growth rate (2008):	6.6%
GDP per capita (2008):	$1,901 U.S.
Goods exported:	Electricity, metal products, timber
Goods imported:	Fuel and lubricants, grains, machinery, vehicles
Trading partners:	Hong Kong, India, Indonesia, Singapore

tains of the Himalaya rise over 24,000 feet (7,320 meters). In addition to its towering peaks, this region includes freezing lakes and huge glaciers. The climate is very cold.

Most of Bhutan is unsuitable for farming, with poor soil and very little flat land. Even so, Bhutan's farmers manage to grow enough food for themselves. Their chief crops include corn, oranges, potatoes, and rice. Farmers in the higher areas herd cattle and yaks.

The Indian government has helped the Bhutanese to modernize their economy by establishing orchards and stock-breeding farms and building hydroelectric power stations. India has also helped Bhutan build roads and train its farmers and workers.

Way of life

Bhutias, people of the Sharchop and Ngalop ethnic groups, make up more than half of the population. Many Bhutias speak Dzongkha, a Tibetan dialect, which is also the country's official language. They practice Lamaism, also known as Tibetan Buddhism.

About a fourth of the Bhutanese people are Nepalese. They speak Nepali and practice Hinduism. In the early 1990's, friction between the Nepali-speaking people of southern Bhutan and the ruling north Bhutanese led to tens of thousands of southern Bhutanese fleeing to refugee camps in Nepal, Assam, and West Bengal.

Taktsang Monastery clings to the side of a sheer cliff in the Himalayan region. It can be reached only by a steep, narrow path. The monastery is known as the "Tiger's Nest" because of Bhutanese legends about the site.

The independent state of Bhutan depends on its huge neighbor to the south, India, to handle its foreign affairs and its defense. India has also helped Bhutan improve its industries and train its workers.

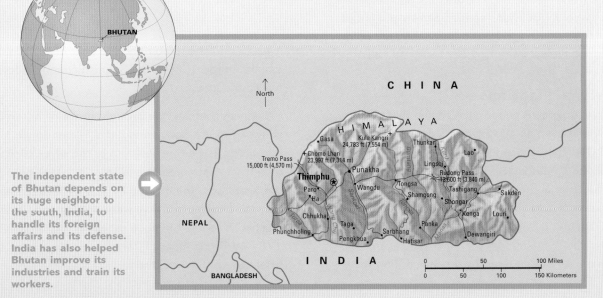

BOLIVIA

Wrapped in woolen ponchos and shawls, three Native American women are ready to begin their daily chores. Their colorful garments stand out boldly against the desolate landscape of Bolivia's Altiplano. A high plateau that lies between two craggy ranges of the Andes Mountains, the Altiplano is home to about 40 percent of the country's population. Many inhabitants of this region live in La Paz, a major urban center and Bolivia's seat of government. *Campesinos*—the poor farmers who make up Bolivia's largest social class—live in the countryside.

Many campesinos are indigenous and wear the traditional clothing of their ancestors. The women dress in full, sweeping skirts that almost touch the ground, and many wear derby hats. The men wear striped ponchos. The campesino customs and way of life are deeply rooted in the past. In many parts of Bolivia, the Indians live much as their ancestors did before the Spaniards conquered them in the 1530's.

A much smaller class of wealthy Bolivians, called the *elite,* live in city apartments or in majestic Spanish-style houses. Wealth has remained in these elite families for generations. Middle class Bolivians resemble the elite but live less luxuriously. Middle class professionals in Bolivia include government officials, doctors, and lawyers. Most wealthy Bolivians speak Spanish and are of European or *mestizo* (mixed European and indigenous) ethnicity.

Early civilizations

Native Americans lived in the region that is now Bolivia as long as 10,000 years ago. About A.D. 100, the Tiahuanaco Indians developed a major civilization near Lake Titicaca. They built gigantic monuments and carved statues out of stone. The Tiahuanaco civilization declined during the 1200's, and by the late 1300's, a warlike tribe called the Aymara Indians controlled the area. The Aymara were conquered by the Inca during the

La Paz is one of the two national capitals of Bolivia. The city sits in a deep valley high in the snow-covered Andes Mountains. The city is the highest capital in the world.

1400's, and the region became part of the great Inca empire. In the 1530's, the Spanish conquistadors defeated the Inca.

Spanish colonial rule brought years of misery and suffering to the Bolivian Indians. Forced to work on the *haciendas* (large ranches) and in the silver mines, many died from mistreatment and from diseases brought by the Spaniards. The Indians frequently revolted against Spanish tyranny, but they were quickly and brutally crushed. The most famous rebellion was organized in 1781 by Tupac Amaru II, who would have been the official heir to the Inca throne.

In the late 1800's, the world price of silver increased greatly, and large deposits of tin were discovered in Bolivia. The export of these minerals became highly important to Bolivia's economy. Political parties representing the interests of the mine owners grew more and more powerful. They controlled Bolivia until the 1930's and helped the country achieve greater political stability. Bolivia's presidents during this time devoted much effort to promoting mining and the building of railroads.

After the mid-1930's, control of Bolivia's government changed hands frequently for most of the rest of the 1900's. Several Bolivian presidents were military leaders whose governments violated the civil rights of the people. The nation's frequent wars and revolutions hampered its economic growth, and today Bolivia has one of the lowest standards of living in the Western Hemisphere.

Native American women sell their produce at an open air market in La Paz. Native Americans have lived in what is now Bolivia for thousands of years.

Bolívar and liberation

In 1824, the great Venezuelan general Simón Bolívar sent Antonio José de Sucre to free Bolivia from Spanish rule. Sucre's forces defeated the Spaniards in 1825, and Bolivia declared its independence. Sucre became the country's first president.

Unfortunately for Bolivia, the tyranny of the Spaniards was replaced by chaos and violence during the years following independence. Many of Bolivia's leaders showed far more concern for their own political power and wealth than for the needs of the people.

In addition, Bolivia lost more than half of its territory in disputes with neighboring countries. In the War of the Pacific (1879–1883), Chile seized Bolivia's nitrate-rich land along the Pacific Ocean. Bolivia has been completely landlocked ever since.

BOLIVIA TODAY

Since gaining independence from Spain in 1825, Bolivia has experienced scores of successful and unsuccessful attempts to overthrow its government. Until the 1950's, Bolivia's people had little political freedom and almost no possibility of social advancement.

Revolution and reform

When the world price of silver shot up in the late 1800's, profits from Bolivia's silver mines helped boost the economy. During the same period, large deposits of tin were discovered. Political parties representing the mine owners came to power and managed to maintain political stability until the Chaco War broke out in 1932.

The Chaco War erupted between Bolivia and Paraguay over ownership of the Gran Chaco, a lowland plain bordering both countries. Bolivia lost the war in 1935, and Paraguay won the Gran Chaco. Tremendous political disorder followed this defeat. During this period, tin miners formed unions and went out on strike for better wages and working conditions.

In 1952, the *Movimiento Nacionalista Revolucionario* (National Revolutionary Movement), a political party supported by the tin miners, overthrew the military government then in power. Victor Paz Estenssoro, an economist and party leader, became president. The new government nationalized the tin mines and introduced land reforms, breaking up the vast estates of the landowners and dividing them among poor Indian farmers. But perhaps most important, all adult Bolivians gained the right to vote, marking the first time the Indian population was included in the political system.

Return to military government

The period of reform came to an end when military dictators regained control of the government. From 1964 to 1982, one violent coup succeeded another. The dictators prohibited any opposition to their policies and imprisoned or killed their enemies. In 1967, Che Guevara, a revolutionary famous for his activities in Cuba, tried to organize a revolt against military rule, but he was captured and executed.

In 1980, Bolivia held an election for a civilian government, but military leaders once again seized power be-

FACTS

● Official name:	República de Bolivia (Republic of Bolivia)
● Capital:	Sucre (legal capital and seat of judiciary), La Paz (seat of government)
● Terrain:	Rugged Andes Mountains with a highland plateau (Altiplano), hills, lowland plains of the Amazon Basin
● Area:	424,165 mi² (1,098,581 km²)
● Climate:	Varies with altitude; humid and tropical to cold and semiarid
● Main rivers:	Grande, Mamoré, Beni
● Highest elevation:	Nevado Sajama, 21,463 ft (6,542 m)
● Lowest elevation:	Near Fortaleza, 300 ft (90 m)
● Form of government:	Republic
● Head of state:	President
● Head of government:	President
● Administrative areas:	9 departmentos (departments)
● Legislature:	Congreso Nacional (National Congress) consisting of Cámara de Senadores (Chamber of Senators) with 27 members serving five-year terms and Cámara de Diputados (Chamber of Deputies) with 130 members serving five-year terms
● Court system:	Corte Suprema (Supreme Court)
● Armed forces:	46,100 troops
● National holiday:	Independence Day - August 6 (1825)
● Estimated 2010 population:	10,040,000
● Population density:	24 persons per mi² (9 per km²)
● Population distribution:	65% urban, 35% rural
● Life expectancy in years:	Male, 64; female, 68
● Doctors per 1,000 people:	1.2
● Birth rate per 1,000:	27
● Death rate per 1,000:	8
● Infant mortality:	48 deaths per 1,000 live births
● Age structure:	0-14: 37%; 15-64: 58%; 65 and over: 5%
● Internet users per 100 people:	11
● Internet code:	.bo
● Languages spoken:	Spanish, Quechua, Aymara (all official)
● Religions:	Roman Catholic 95%, Protestant 5%
● Currency:	Boliviano
● Gross domestic product (GDP) in 2008:	$17.68 billion U.S.
● Real annual growth rate (2008):	5.6%
● GDP per capita (2008):	$1,810 U.S.
● Goods exported:	Mostly: natural gas Also: soybeans and soy products, sugar, tin, zinc
● Goods imported:	Food, heavy machinery, petroleum products, transportation equipment
● Trading partners:	Mostly: Brazil Also: Argentina, Chile, Japan, United States

Bolivia lies south of the equator near the center of South America. Despite its wealth of natural resources, Bolivia's economic progress has been hampered by political instability and a poor transportation system.

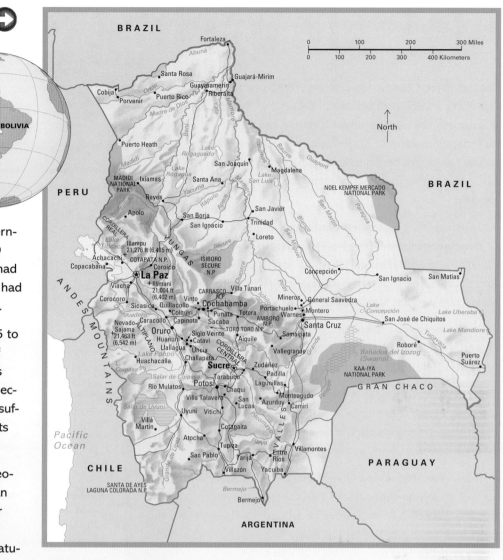

fore the elected government could take office. Then, in 1982, the military relaxed its grip and allowed a return to civilian government. The Congress elected in 1980 chose Siles Zuazo as president. He had been president 22 years earlier and had been elected by the people in 1980.

In presidential elections from 1985 to 2002, no candidate got a majority of the popular vote. Therefore, Congress chose the president following each election. Bolivia's economy continued to suffer inflation, while floods and droughts led to food shortages.

In September 2003, dozens of people, many of whom were of American Indian descent, died in protests after President Sánchez de Lozada announced a plan to export Bolivian natural gas to the United States and Mexico. The protesters demanded social reforms, feared foreign exploitation of Bolivia's natural resources, and opposed plans to destroy coca crops as part of a war on illegal drugs. In October, Sánchez de Lozada resigned and fled the country to escape trial.

Bolivians elected Evo Morales president in December 2005. Morales, an Aymara Indian and Bolivia's first indigenous president, led a party called *Movimiento a Socialismo* (Movement Toward Socialism). Morales was reelected in 2009.

Also in 2009, a majority of Bolivians voted for a new constitution. Proponents hoped the new constitution would give the poor a greater voice in government.

The Presidential Palace is the official residence of the president of Bolivia. The palace stands on one side of Murillo Square in La Paz. The square is named for Pedro Domingo Murillo, a Bolivian patriot.

BOSNIA-HERZEGOVINA

The republic of Bosnia-Herzegovina *(BOZ nee uh HURT suh goh VEE nuh)* lies in the Balkan Peninsula in southeastern Europe. It was formerly one of the six federal republics of Yugoslavia. In 1991, the Yugoslav federation began to break apart, and in 1992 Bosnia-Herzegovina became an independent republic.

Geographically, Bosnia-Herzegovina is made up of two regions. Bosnia, the northern part, is a mountainous land covered with thick forests, while Herzegovina, the southern part, consists mainly of rolling hills and flat farmland.

Bosnia-Herzegovina remains primarily a rural country with few cities and towns. Its chief industries produce electrical appliances and textiles.

Bosnia-Herzegovina is a culturally diverse region, blending Islamic, Christian, Central European, and Mediterranean traditions. The languages spoken are Bosnian, Croatian, and Serbian.

The region that is now Bosnia-Herzegovina was settled by Illyrian tribes about 3,000 years ago. It became part of a Roman province around 11 B.C. Slavs immigrated to the region in the late A.D. 500's and the 600's. From the 900's to the 1100's, control passed between the Byzantine Empire, Croatia, and Serbia. Hungary's king claimed authority over most of Bosnia from the 1100's to the 1400's, but local nobles were able to act independently much of the time. Hum (now Herzegovina) fell under Serbian or Hungarian rule from the 1100's until 1326. Bosnia controlled it from 1326 until 1448, when its local ruler declared independence.

Bosnia-Herzegovina was part of the Ottoman Empire from the late 1400's until Austria-Hungary gained control in 1878. In 1914, a Serbian patriot from Bosnia assassinated the heir to the throne of Austria-Hungary in the city of Sarajevo. The assassination touched off World War I (1914-1918). After the war ended, Bosnia-Herzegovina became part of the Kingdom of the Serbs, Croats, and Slovenes, later renamed Yugoslavia.

FACTS

Official name:	Bosna I Hercegovina (Bosnia and Herzegovina)
Capital:	Sarajevo
Terrain:	Mountains and valleys
Area:	19,767 mi² (51,197 km²)
Climate:	Hot summers and cold winters; areas of high elevation have short, cool summers and long, severe winters; mild, rainy winters along coast
Main rivers:	Bosna, Drina, Neretva, Sava, Una
Highest elevation:	Mount Maglic, 7,828 ft (2,386 m)
Lowest elevation:	Adriatic Sea, sea level
Form of government:	Emerging federal democratic republic
Head of state:	Chairman of the presidency
Head of government:	Chairman of the Council of Ministers
Administrative areas:	2 first-order administrative divisions and one internationally supervised district
Legislature:	Skupstina (Parliamentary Assembly) consisting of the Predstavnicki Dom (National House of Representatives) with 42 members serving four-year terms and the Dom Naroda (House of Peoples) with 15 members serving four-year terms
Court system:	Constitutional Court
Armed forces:	8,500 troops
National holiday:	National Day - November 25 (1943)
Estimated 2010 population:	3,968,000
Population density:	201 persons per mi² (78 per km²)
Population distribution:	54% rural, 46% urban
Life expectancy in years:	Male, 73; female, 80
Doctors per 1,000 people:	1.4
Birth rate per 1,000:	9
Death rate per 1,000:	9
Infant mortality:	9 deaths per 1,000 live births
Age structure:	0-14: 17%; 15-64: 69%; 65 and over: 14%
Internet users per 100 people:	35
Internet code:	.ba
Languages spoken:	Bosnian, Croatian, Serbian
Religions:	Muslim 40%, Orthodox 31%, Roman Catholic 15%, Protestant 4%, other 10%
Currency:	Convertible marka
Gross domestic product (GDP) in 2008:	$18.76 billion U.S.
Real annual growth rate (2008):	5.6%
GDP per capita (2008):	$4,782 U.S.
Goods exported:	Aluminum and other metals, machinery, wood products
Goods imported:	Chemicals, food, fuels, machinery, transportation equipment
Trading partners:	Austria, Croatia, Germany, Italy, Slovenia

Formerly a federal republic of Yugoslavia, Bosnia-Herzegovina became an independent republic in 1992.

Mostar is one of the largest cities in Bosnia. It sits along the Neretva River. The Old Bridge, built in the 1500's during the Ottoman era, is a symbol of the city.

In 1946, Bosnia-Herzegovina became one of the six republics of the Federal People's Republic of Yugoslavia. In March 1992, a majority of Bosniaks (sometimes called Bosnian Muslims) and ethnic Croats in Bosnia-Herzegovina voted for independence from Yugoslavia in a referendum that was boycotted by the Serbs.

Fighting broke out between the Serbs, who claimed part of the republic, and the Bosniaks and Croats. The Serbs, who wished to "cleanse" the region of all non-Serbs, soon gained control of more than two-thirds of Bosnia-Herzegovina. Much of the fighting centered around the capital city of Sarajevo.

In June 1994, a United Nations (UN) commission accused Bosnian Serbs of a campaign of genocide against Bosnian Croats and Bosniaks. In November, a Bosniak army offensive drove Serbs off land they held in northwest Bosnia-Herzegovina. In retaliation, the Serbs attacked Bihac, a Bosniak stronghold.

In 1995, the presidents of Bosnia-Herzegovina, Croatia, and Serbia agreed on a peace plan. Under the plan, Bosnia-Herzegovina would keep its borders but split into two substates, one dominated by Bosnian Serbs and one by the Bosniak-Croat federation. In 1996, Bosniak, Croat, and Serb forces withdrew from the zones of separation established by the peace agreement. The exchange of territory was marked by violence.

In the early 2000's, Bosnia struggled to recover from the damage done by the war. High unemployment and organized crime added to the problems. Nationalist parties, which sought power for individual ethnic groups, and reformist parties, which sought to unite the country's ethnic groups, traded control of the government. After parliamentary elections in October 2010, Bosnia spent 15 months without a national government because the parties could not agree. A new government was finally formed in December 2011.

BOTSWANA

Botswana lies far from the sea in the center of southern Africa. More than 100 years ago, after the local people asked for British protection from white South African settlers, it was governed by the United Kingdom and called the Bechuanaland Protectorate. After repeatedly refusing South Africa's requests to turn the protectorate over to South African rule, the United Kingdom granted the region its independence in 1966, and the Republic of Botswana was born. Today, Botswana is a democratic republic.

Government and people

Citizens who are at least 18 years old vote for candidates from several different parties to represent them as members of the National Assembly. The Assembly members then choose a president and four additional Assembly members. The president selects a cabinet to help run the government. In addition, the House of Chiefs, made up of the leaders of Botswana's major ethnic groups, advises the government.

The majority of Botswanans belong to the Tswana ethnicity, which is divided into eight main groups. Most of the Tswana live in large rural villages and farm or herd for a living. Their main food crops are corn, millet, and sorghum.

Botswana also has about 30,000 San, or Bushmen, an African people whose ancestors have lived in the region since prehistoric times. Sometime before A.D. 1000, the San were pushed from fertile eastern Botswana into the Kalahari region by the Tswana, who had migrated from the north. A few San still live in the Kalahari, gathering food and hunting as their ancestors did, but many have been forced into settlements. Some work on cattle farms.

Fish from the Okavango River are a welcome addition to the diet of this Tswana villager.

A number of whites also live in Botswana. Some own ranches, while others are technicians or managers in industry, business, or government. Generally, the whites earn more money and have a higher standard of living than black Botswanans.

FACTS

Official name:	Republic of Botswana
Capital:	Gaborone
Terrain:	Predominantly flat to gently rolling tableland; Kalahari Desert in southwest
Area:	224,607 mi² (581,730 km²)
Climate:	Semiarid; warm winters and hot summers
Main rivers:	Okavango, Limpopo, Shashe
Highest elevation:	Otse Mountain, 4,886 ft (1,489 m)
Lowest elevation:	Junction of the Limpopo and Shashe Rivers, 1,684 ft (513 m)
Form of government:	Republic
Head of state:	President
Head of government:	President
Administrative areas:	9 districts, 5 town councils
Legislature:	Parliament consisting of the House of Chiefs, which is an advisory 15-member body, and the National Assembly with 61 members serving five-year terms
Court system:	High Court; Court of Appeal; Magistrates' courts
Armed forces:	9,000 troops
National holiday:	Independence Day (Botswana Day) - September 30 (1966)
Estimated 2010 population:	1,893,000
Population density:	8 persons per mi² (3 per km²)
Population distribution:	58% urban, 42% rural
Life expectancy in years:	Male, 56; female, 56
Doctors per 1,000 people:	0.4
Birth rate per 1,000:	24
Death rate per 1,000:	14
Infant mortality:	33 deaths per 1,000 live births
Age structure:	0-14: 35%; 15-64: 62%; 65 and over: 3%
Internet users per 100 people:	5
Internet code:	.bw
Languages spoken:	Tswana, Kalanga, Sekgalagadi, English (official)
Religions:	Christian 71.6%, Badimo 6%, other 22.4%
Currency:	Pula
Gross domestic product (GDP) in 2008:	$13.41 billion U.S.
Real annual growth rate (2008):	3.2%
GDP per capita (2008):	$7,630 U.S.
Goods exported:	Mostly: diamonds Also: copper, meat, textiles
Goods imported:	Chemicals, food, fuel, machinery, transportation equipment
Trading partners:	South Africa, United Kingdom, United States, Zimbabwe

Economy

Botswana is a poor country. From the late 1960's through the mid-1990's, the country's economy developed rapidly. In the 1990's and early 2000's, however, economic growth slowed due to the rapid spread of AIDS.

Botswana's two chief industries are mining and the raising of livestock, especially cattle. Copper, diamond, and nickel deposits discovered in the late 1960's and the 1970's are now being developed. The country also has deposits of coal and cobalt.

The manufacturing industry in Botswana is developing through government encouragement of private enterprise and foreign investment. However, unemployment remains a major problem in the country. Thousands of Botswanans, most of them young men, work in South Africa for several months a year. While this arrangement brings badly needed money into Botswana, it separates families and causes other social problems.

Poor housing is also a major concern. Thousands of rural Botswanans move to the cities each year, hoping to find work and a better life. Unfortunately, many must live in crowded slums, especially in mining towns such as Orapa and Selebi-Pikwe.

Currently, the country's economy depends heavily on South African investments, markets, and technical skills. In addition, nearly all of Botswana's imports and exports travel on a railroad that runs through South Africa to the sea.

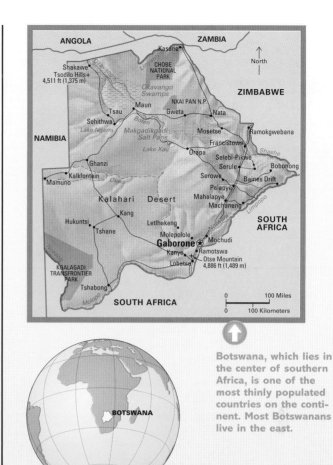

Botswana, which lies in the center of southern Africa, is one of the most thinly populated countries on the continent. Most Botswanans live in the east.

A herd of elephants grazes in Chobe National Park, one of several wildlife reserves in Botswana, where these threatened animals are protected from ivory poachers. The Kalahari Desert covers the center and southwest of the country. The Okavango River forms a vast marshland in the northwest.

BRAZIL

As the sun rises on Rio de Janeiro, another day begins in one of the world's most beautiful cities. Soon, sunbathers will flock to the world-famous Copacabana and Ipanema beaches. In the business district—also the center of Brazilian finance—stock market traders are already at their computer terminals. Along the city boulevards, shopkeepers prepare to open their elegant boutiques.

Bustling with activity by day, Rio de Janeiro quickens its pace even more after the sun sets. Festive partymakers crowd the city's exciting nightclubs, and colorful street festivals light up the darkness.

With its breathtaking setting, fascinating sights, and unique character, Rio de Janeiro is like Brazil itself, seemingly larger than life. Brazil—sprawled across almost half a continent and charged with an energy all its own—is a land that captures the visitor's imagination.

Brazil is South America's largest country in area, and it has about as many people as all the other countries combined. The course of its mighty Amazon River is longer than the highway route between New York City and San Francisco. Most of northern Brazil is covered by the largest tropical rain forest on Earth. In other parts of the country, miles and miles of dry, grassy plains stretch across the countryside.

Cloud-capped mountains rise north of Brazil's forests and border the Atlantic Ocean in the southeast. The low plateaus of central and southern Brazil have fertile farmlands and lush grazing areas. Broad, white beaches line glistening seashores on the nation's Atlantic coast.

The forests, rivers, and mountains of Brazil have restricted inland travel, and the country's vast interior remains largely undeveloped. About 80 percent of all Brazilians live within 200 miles (320 kilometers) of the Atlantic coast. One of the largest cities in Brazil's interior is Brasília, the nation's capital. It was built about 600 miles (970 kilometers) from the coast to help draw Brazilians inland.

Brazil is a land of many contrasts. Upper-class landowners enjoy the luxuries of modern life, while the nation's rural poor live in small huts with dirt floors. Brazil is a major industrial nation and a tremendous economic power, yet it has a staggering amount of debt. Every day, more of its precious rain forest is cleared for development, threatening not only the survival of animals and plant species, but also the well-being of our entire planet.

Despite its problems, however, Brazil has a spirit of warmth and liveliness unequaled on the South American continent. From the sandy beaches on its east coast to the tropical rain forests of the interior, this huge country offers much to see, explore, and appreciate.

BRAZIL TODAY

Brazil's great natural resources have made the country a potent and growing economic power. The nation's fertile farmland yields a huge coffee and banana crop, and its forests provide timber, nuts, and other products. Brazil's mines produce large quantities of iron ore, manganese, and other minerals needed by industry, while its rivers generate huge amounts of electricity.

Despite rapid industrial growth in the mid-1900's, Brazil still faces serious economic problems. Chief among these problems are widespread poverty and unemployment. The military regime that ruled Brazil from the mid-1960's to the mid-1980's borrowed huge amounts of money from other nations to finance industrial development. As a result, Brazil's foreign debt grew to astronomical proportions. Brazil returned to civilian government in 1985.

The urban poor

Although the middle class has grown significantly, a huge gap still exists between the few who are enormously rich and the great mass of poor citizens. This extremely uneven distribution of wealth has been a serious obstacle to social progress. Part of the problem stems from land policies dating from the 1600's, when wealthy plantation owners acquired huge tracts of land.

Rural people with no land to farm continue to migrate to the cities, hoping to find work and a better life. But because many of these people are unskilled and uneducated, they can get only low-paying jobs—if they can find work at all. With little or no income, these rural migrants are forced to live in urban slums known as *favelas*.

Houses in the favelas are generally shabby shacks made of cardboard, metal, or wood. Because of the poor sanitation, many people suffer from diseases. Many parents abandon their children because they cannot afford to feed them. Millions of Brazilian children live on the streets—begging, stealing, or working long hours just to survive.

FACTS

Official name:	Republica Federativa do Brasil (Federative Republic of Brazil)
Capital:	Brasília
Terrain:	Mostly flat to rolling lowlands in north; some plains, hills, mountains, and narrow coastal belt
Area:	3,287,613 mi² (8,514,877 km²)
Climate:	Mostly tropical, but temperate in south
Main rivers:	Amazon, São Francisco, Araguaia, Paraná, Parnaíba
Highest elevation:	Pico da Neblina, 9,888 ft (3,014 m)
Lowest elevation:	Atlantic Ocean, sea level
Form of government:	Federal republic
Head of state:	President
Head of government:	President
Administrative areas:	26 estados (states), 1 distrito federal (federal district)
Legislature:	Congresso Nacional (National Congress) consisting of the Senado Federal (Federal Senate) with 81 members serving eight-year terms and the Camara dos Deputados (Chamber of Deputies) with 513 members serving four-year terms
Court system:	Supreme Federal Tribunal
Armed forces:	326,400 troops
National holiday:	Independence Day - September 7 (1822)
Estimated 2010 population:	199,132,000
Population density:	61 persons per mi² (23 per km²)
Population distribution:	85% urban, 15% rural
Life expectancy in years:	Male, 69; female, 75
Doctors per 1,000 people:	1.2
Birth rate per 1,000:	19
Death rate per 1,000:	6
Infant mortality:	23 deaths per 1,000 live births
Age structure:	0-14: 27%; 15-64: 67%; 65 and over: 6%
Internet users per 100 people:	36
Internet code:	.br
Languages spoken:	Portuguese (official), Spanish, German, Italian, Japanese, English
Religions:	Roman Catholic (nominal) 73.6%, Protestant 15.4%, other 11%
Currency:	Real
Gross domestic product (GDP) in 2008:	$1.617 trillion U.S.
Real annual growth rate (2008):	5.2%
GDP per capita (2008):	$8,354 U.S.
Goods exported:	Airplanes, cars, coffee, iron ore, meat, shoes, soybeans and soy meal, steel, sugar
Goods imported:	Chemicals, machinery, petroleum, pharmaceuticals, transportation equipment
Trading partners:	Argentina, China, France, Germany, Italy, Japan, Netherlands, United States

COLOMBIA

VENEZUELA
Mt. Roraima 9,094 ft (2,772 m)
MONTE RORAIMA N.P.

SURINAME

FRENCH GUIANA
Cape Orange

Guiana Highlands
Boa Vista ★
Caracaraí

GUYANA

CABO ORANGE N.P.
Maracá I.
Amapá

North Atlantic Ocean

ECUADOR

Pico da Neblina 9,888 ft (3,014 m)
PICO DA NEBLINA NATIONAL PARK

SERRA DA MOCIDADE N.P.
Roraima

VIRUÁ N.P.

MONTANHAS DO TUMUCUMAQUE N.P.

Amapá
Macapá
Mazagão

Marajó Estuary

0° Equator

JAÚ N.P.

Monte Dourado
Monte Alegre
Oriximiná

Marajó I.
Bragança
Belém
Castanhal

São Marcos Bay

São Luís
Cametá

LENÇÓIS MARANHENSES N.P.
JERICOACOARA N.P.

Manaus
Manacapuru
Itacoatiara
Parintins
Tupinambarana I.

Santarém
AMAZONIA NATIONAL PARK

Pará
Altamira

Rocas Atol

Tefé
Coari

Selva (Tropical rain forest)

Marabá

Bacabal
Codó
Coxias
Timon
Teresina

Sobral
Caucaia

UBAJARA N.P.
Ceará

Fortaleza
Mossoró
Cape São Roque

Benjamin Constant

Amazonas

TRANS-AMAZON HIGHWAY

Carajás
Serra Pelada

Imperatriz

Maranhão

Boa Esperança Res.

Floriano
Juázeiro do Norte

Crato

Rio Grande do Norte
Currais Novos
Paraíba

Natal
João Pessoa

Cruzeiro do Sul
SERRA DO DIVISOR N.P.
Feijó
Acre

Humaitá

Conceição do Araguaia

ARAGUAIA NATIONAL PARK

SERRA DAS CONFUSÕES NATIONAL PARK
Piauí

NASCENTES DO RIO PARNAÍBA N.P.

Petrolina
Juázeiro

Pernambuco
SERRA DA CAPIVARA N.P.
Campina Grande
Caruaru

Recife
Garanhuns
Maceió
Alagoas

Pôrto Velho

Guajará-Mirim

Ji-Paraná

Serra Formosa

Palmas
Pôrto Nacional
Dianópolis

Sertão

Bôca do Acre
Rio Branco

PACAÁS NOVOS N.P.
Rondônia

SERRA DA CUTIA N.P.

Serra do Tombador

Tocantins

Barreiras

Bahia

Feira de Santana

Aracaju
Sergipe
Estância

SERRA DOS PARECIS

Porangatu

CHAPADA DOS VEADEIROS N.P.

Alagoinhas
Camaçari
Salvador

Mato Grosso

Mato Grosso Plateau

CHAPADA DOS GUIMARÃES N.P.
Várzea Grande
Cuiabá

Barra do Garças

Distrito Federal
Brasília ⊛
Formosa

Brazilian Highlands

Bom Jesus da Lapa

Todos os Santos Bay

Cáceres

Rondonópolis

Anápolis
Goiânia

Goiás

Montes Claros
Minas Gerais

Serra do Espinhaço

Itabuna
Vitória da Conquista
Ilhéus

DESCOBRIMENTO N.P.
PAU BRASIL N.P.
MONTE PASCOAL N.P.

BOLIVIA

PANTANAL MATO GROSSENSE N.P.

Corumbá

Taquari

EMAS N.P.
Jataí

Itumbiara

Uberlândia

Patos de Minas

SEMPRE-VIVAS N.P.

Caravelas

MARINHO DOS ABROLHOS N.P.

North

Aquidauana

SERRA DA BODOQUENA N.P.

Paraguay

Mato Grosso do Sul

Campo Grande

Sete Lagoas
SERRA DA CANASTRA N.P.

Belo Horizonte
Divinópolis

Governador Valadares
Espírito Santo
Linhares

Pico da Bandeira 9,482 ft (2,890 m)
Vitória
Vila Velha

Bela Vista
Ponta Porã

Dourados

Araçatuba

São José do Rio Prêto

Ribeirão Prêto
Barbacena

Volta Redonda
ITATIAIA N.P.

Campos
Cachoeiro de Itapemirim
Nova Friburgo

South Pacific Ocean

PARAGUAY

Presidente Prudente
Maringá

Paraná

ILHA GRANDE N.P.

São Paulo
Claro

Campinas
Sorocaba

Santo André

Nova Iguaçu
Niterói Rio de Janeiro

Rio de Janeiro

Santos

Brazil ranks among the largest, most populated countries in the world. Brazil was named after the brazilwood trees that grow there.

Londrina
Ponta Grossa

São Bernardo do Campo

CHILE

Cascavel

IGUAÇU N.P.

Guarapuava

Curitiba
Paranaguá

South Atlantic Ocean

0 — 500 Miles
0 — 500 Kilometers

Iguaçu Falls (Iguazú Falls)

Chapecó

Santa Catarina
Blumenau

Joinvile
Itajaí
Florianópolis

ARGENTINA

Passo Fundo

Lajes

SÃO JOAQUIM N.P.

Uruguaiana

Santa Maria

Caxias do Sul
Rio Grande do Sul
Canoas

Criciúma
SERRA GERAL N.P.
APARADOS DA SERRA N.P.

Pôrto Alegre
Bagé
Pelotas

LAGOA DO PEIXE N.P.

BRAZIL

URUGUAY

Rio Grande
Mirim Lake

Democratic government

In 1985, Brazil returned to a civilian government after 21 years of military rule. When military leaders first took over the Brazilian government in 1964, the country's economy flourished. But the mid-1970's brought a worldwide business slump, and Brazil's economic growth slowed down. By 1979, the military administration faced such problems as high inflation rates and labor unrest.

In 1985, the electoral college elected a civilian president, Tancredo de Almeida Neves. Neves died soon after, and José Sarney, who had been elected vice president, was named president. A constitutional amendment passed by Congress in 1985 provided for the direct election of future presidents by the people.

In December 1989, the people elected Fernando Collor de Mello president. In 1992, minutes after his Senate impeachment trial began and facing criminal charges for corruption, Collor resigned his post. Itamar Franco became the new president. In 1994, Collor was acquitted of the charges. Fernando Cardoso was elected president in 1995 and reelected in 1998. Cardoso was credited with reducing inflation and the deficit.

Luiz Inácio Lula da Silva, who served as president from 2002 to 2010, promoted policies that addressed poverty. Like Cardoso, he stressed economic growth and social services. In 2011, Dilma Rousseff, who had served as da Silva's chief of staff, became Brazil's first woman president.

HISTORY

In 1493, the year after Christopher Columbus arrived in the New World, Pope Alexander VI drew an imaginary north-south line that divided the lands being explored and claimed by Spanish and Portuguese navigators. Known as the Line of Demarcation, it was intended to establish a boundary that would prevent disputes between Spain and Portugal over these lands. Land to the east of the line was declared Portuguese territory, while land to the west belonged to Spain. In 1494, through the Treaty of Tordesillas, Spain and Portugal moved the line westward to a point about 1,295 miles (2,084 kilometers) west of the Cape Verde Islands, giving Portugal what is now eastern Brazil.

Early settlements

In 1500, Pedro Álvares Cabral landed on the Brazilian coast to claim Portugal's new territory. But it was not until the 1530's that Portuguese colonists began to settle the region and establish huge sugar plantations. During the 1600's and 1700's, many Portuguese settlers migrated to the interior and south of Brazil, where gold and diamonds had been discovered.

As colonists in the north began migrating into Brazil's interior, they crossed the Line of Demarcation. In 1750, Spain and Portugal signed the Treaty of Madrid, which gave Portugal almost all of what is now Brazil.

The splendor of baroque design lights up the church of Nossa Senhora do Pilar in the town of Ouro Prêto. According to local legend, the craftsmen who decorated the 300-year-old church mixed 182 pounds (400 kilograms) of gold dust in their paints.

A new nation

In 1807, France invaded Portugal, and Prince John, Portugal's ruler, fled to Rio de Janeiro. In 1808, Rio became the capital of the Portuguese Empire, and Prince John raised Brazil to the status of a kingdom. When the royal family returned to Portugal in 1821, Prince John left his son Pedro to rule the new kingdom.

On Sept. 7, 1822, Pedro declared Brazil's independence from Portugal. A few months later, he was named emperor. But Pedro was a harsh ruler, and he became so unpopular that he was forced to resign in 1831. He left his throne to his 5-year-old son, Pedro II.

The Indians tried to defend their native land against the Portuguese colonists, and many were killed in battle. Many others were forced to work as slaves on the plantations. When the Indian population was devastated by European diseases, the plantation owners replaced them with black slaves from Africa.

TIMELINE

1500	Pedro Álvares Cabral lands on Brazil's east coast and claims the country for Portugal.
1530's	Portuguese colonists settle in northeast and southern Brazil.
1549	Salvador is founded as the capital of the Brazilian colony.
1565	Portuguese establish a fort, which grows into the city of Rio de Janeiro.
1630	The Dutch invade Brazil.
1654	Portuguese drive the Dutch out of Brazil.
1700's	Portuguese colonists travel to Brazil's interior in search of gold and diamonds.
1750	Portugal and Spain sign the Treaty of Madrid.
1763	Rio de Janeiro becomes the capital of Brazil.
1808–1821	Portuguese royal family rules Portugal and Brazil from Rio de Janeiro.
1822	Brazil declares its independence.
1831	Pedro I is forced to give up his throne, and 5-year-old Pedro II becomes emperor of Brazil.
Mid-1800's	Thousands of European immigrants settle in southern Brazil.
1888	Pedro II abolishes slavery and frees about 750,000 slaves.
1889	Brazil becomes a republic.
Early 1900's	Coffee becomes Brazil's chief export, bringing great wealth to the nation.
1917	Brazil joins the Allies in World War I (1914–1918).
1930	Military officers appoint Getúlio Vargas president.
1937	Vargas begins rule as dictator.
1942	Brazil joins the Allies in World War II (1939–1945).
1945	Brazil joins the United Nations.
1946	A new Constitution restores civil rights to the people.
1955	Juscelino Kubitschek is elected president.
1956	Construction of Brasília, the new capital, begins.
1960	The government moves from Rio de Janeiro to Brasília.
1964	Military officers take control of the government.
1974	General Ernesto Geisel becomes president.
1979	General João Baptista Figueiredo succeeds Geisel as president.
1985	Military rule ends, and José Sarney becomes president.
2007-2008	Petrobras, Brazil's state-controlled oil company, announces that it has discovered huge oil fields off the country's southeast coast.
2011	Dilma Rousseff becomes Brazil's first woman president.

Pedro Álvares Cabral (1467?–1528?) Navigator

Pelé (1940–) Soccer player

Luiz Inacio Lula da Silva (1945–) President, 2003-2010

During the long reign of Pedro II, Brazil enjoyed a period of rapid development. The government built railways, telegraph systems, and schools. The growth of industry attracted thousands of European immigrants, who settled in southern Brazil where coffee growing spread rapidly. In addition, the worldwide demand for rubber products led to the development of the Amazon Region's vast natural rubber resources.

In 1888, Pedro II abolished slavery but refused to pay the plantation owners for their slaves. In 1889, the angry plantation owners supported the nation's military officers in removing Pedro from the throne and declaring Brazil a republic. General Manoel Deodoro da Fonseca was elected the first president.

But Deodoro and some of Brazil's other early presidents ruled the country as dictators. Getúlio Vargas, who became president in 1930, was hailed as a national hero when he increased wages and shortened work hours. But in 1937, he prepared a new Constitution that allowed him to censor the press, ban political parties, and take over Brazil's labor unions. Military officers removed Vargas from office in 1945, and the following year a new Constitution restored the people's civil rights.

Vargas was elected president again in 1950, but military officers overthrew his government four years later. In 1955, Juscelino Kubitschek was elected president. Political tension increased in the early 1960's, and military officers again took over the government in 1964.

Military rule ended in 1985. The electoral college elected a civilian president, and a 1985 constitutional amendment provided for the direct election of future presidents by the people.

The magnificent Opera House of Manaus opened in 1896.

PEOPLE

The population of Brazil ranks among the world's largest. Its approximately 200 million people live in a *melting pot* society, where individuals from many different ethnic groups live together in relative harmony. While racial discrimination is far less widespread in Brazil than in many other countries, Brazilians of European descent usually have better educational opportunities and thus hold higher-level jobs in government and industry compared to Brazilians of non-European background.

Only about 8 percent of Brazil's people live in the Amazon Region, which is larger than the United States west of the Mississippi, but mainly covered by thick forests. Most Brazilians live in the cities and urbanized regions along the Atlantic coastal zone. These cities suffer from overcrowding because Brazil has one of the fastest-growing populations in the Western Hemisphere. Many people continue to migrate to the cities from rural areas in search of work.

Brazil's government has made many attempts to redistribute the population. In 1960, the capital was moved from the coastal city of Rio de Janeiro to Brasília, about 600 miles (970 kilometers) inland on the central plateau. During the 1970's, the government began to offer free land to people willing to settle in the Amazon Region.

Roman Catholics make up more than 70 percent of Brazil's population. More Catholics live in Brazil than in any other country. Protestants make up about 15 percent of Brazil's population.

A Brazilian mulatto has mixed African and European descent. The country's many ethnic groups are unified by a shared religion and a common language. More than 70 percent of the people are Roman Catholics, and almost all Brazilians speak Portuguese, the country's official language.

Ethnic groups

About 1 million to 5 million Indians were living in Brazil when the first Europeans arrived in 1500, but the Indian population today is only about 700,000. Most Indians live in the forests of the Amazon Region and speak traditional Indian languages. Numerous Amazonian tribes live in the forests, in settlements that are seldom larger than 200 people.

Racial discrimination in Brazil is less widespread than in many other countries with people of several ethnic groups. But Brazilians of European descent have had better educational opportunities. As a result, they hold most of the higher jobs in government and industry. Many Brazilians of non-European descent have excelled in the arts, entertainment, and sports.

In addition to the Indians, Brazil has three main ethnic groups—people of European origins, people of African descent, and people of mixed ancestry. According to the Brazilian government, people of European origins make up about 55

Followers of Macumba, a local religion that combines African spiritual beliefs with Roman Catholicism, prepare an offering to Iemanjá, the goddess of the sea. Many Brazilians who consider themselves Roman Catholics also worship the gods and goddesses of certain African religions.

percent of the population. However, the government considers many light-skinned people of mixed ancestry as whites.

The white population in the north traces its ancestry to the Portuguese plantation owners who settled there in the 1600's and 1700's. The northeast also contains a distinctive group of blond, blue-eyed Brazilians, the descendants of Dutch colonists who held what is now the state of Pernambuco between 1630 and 1654.

Most European immigrants came to Brazil after the country declared its independence in 1822. People from Germany, Italy, Spain, and Portugal flocked to southeastern Brazil to work in the rapidly growing coffee industry. Today, Brazil has immigrants from dozens of nations. The largest groups include Italians, Portuguese, Spaniards, Japanese, Germans, Poles, and people from the Middle East and the former Soviet republics.

Caboclos and mulattoes

Caboclos (people of mixed European and Indian ancestry) and *mulattoes* (people of mixed African and European ancestry) make up nearly 40 percent of Brazil's population. People of African descent make up about 6 percent of the population.

Beginning in the mid-1500's, people were brought from Africa to Brazil as slaves. Many slaves were sold to sugar cane growers in the northeast. By about 1800, there were so many slaves in Brazil that they made up more than half the population. Many of the descendants of these African slaves still live in the coastal towns and cities, especially in the northeast.

A young lace maker works at her spindle, while colorful shawls and other items are displayed for prospective buyers. The making of renda (lace) is one of the traditional crafts still practiced in the states of Bahia and Ceará.

A fan waves the flag of Brazil in Rio de Janeiro before the start of the Brazilian Grand Prix, one of a series of international races for the world's Formula One championship. Automobile racing is a popular sport in Brazil.

ENVIRONMENT

From the lush vegetation of the tropical rain forests to the arid Northeast and fertile Mato Grosso, Brazil's landscape is beautiful and dramatic. Because all but the southernmost part of Brazil lies in the tropics, the climate is warm to hot the year around, with plenty of rainfall.

The Amazon Region

Extending across most of northern Brazil, the Amazon Region consists mostly of lowlands covered by jungle and tropical rain forest called *selva*. The region, which takes its name from the Amazon River that flows through it, also has two mountain areas—the Guiana Highlands in the far north and the Brazilian Highlands in the south.

The selva lies around the Amazon River and its tributaries, and it contains a tremendous variety of plant and animal life. More than 1,500 kinds of birds and more than 40,000 varieties of plants live in the forests. Scientists have found over 3,000 kinds of trees in 1 square mile (1.6 square kilometers) of the selva, and its animals include many kinds of monkeys as well as anteaters, jaguars, and sloths.

The climate in the western part of the Amazon Region is always hot and humid. Rain falls throughout the year, especially between December and May, for an annual total of about 160 inches (400 centimeters). The eastern part of the region receives less rain—about 40 to 80 inches (100 to 200 centimeters) annually.

The Northeast Region

The Northeast Region consists of the part of Brazil that juts out into the Atlantic Ocean. Although the region occupies less than one-fifth of Brazil's total land area, about 30 percent of its people live there—mainly on the coastal plains, where the fertile red soil drew the first Portuguese colonists to establish sugar cane plantations. Today, farmers still grow sugar cane, as well as cacao and tobacco.

Fantastic rock formations, shaped by the wind and rain over thousands of years, are a feature of Vila Velha Park, in the southern state of Paraná. A human or animal form can be identified in most of the 23 different formations.

A rain forest in the Brazilian Amazon was cut and burned so the land could be turned into a cattle ranch. Cattle ranching is the biggest cause of deforestation in Brazil.

Spectacular Iguaçu Falls on the border between Brazil and Argentina is about 2 miles (3 kilometers) wide. The waters of some 30 rivers and streams plunge 237 feet (72 meters) down the cliffs in 275 separate waterfalls.

Inland from the coastal plains lie the interior backlands, also known as the *sertão*. The sertão consists of plateaus and the hilly sections of the Brazilian Highlands. Agricultural production is low, due to the generally poor soil and the grazing land on the sertão and the variable rainfall, which may cause floods one year and droughts in another.

Central and Southern Plateaus

South of the Amazon Basin and the sertão lie the Central and Southern Plateaus, which include most of the Brazilian Highlands. A steep slope called the Great Escarpment runs along the Atlantic coast on the southeastern ridge of the highlands. This slope has been a partial barrier to the development of Brazil's interior.

The climate of the plateaus is cooler than the Amazon and Northeast regions. Winter frosts often occur in the state of Paraná, and light snow sometimes falls in the state of Santa Catarina. More than half of Brazil's people live in the region of the Central and Southern Plateaus, known as the nation's economic heartland, because of its fertile soil, fine cattle ranches, and rich mines. Farmers grow coffee on large plantations called *fazendas*.

Brazil's geography has strongly influenced the country's pattern of settlement. The hot, humid Amazon rain forest remains largely uninhabited, while the fertile soil and milder climate of southern Brazil have attracted most of the population.

The strong waves of the Atlantic make the coast of Bahia a surfer's paradise. From the sandy white beaches of Rio de Janeiro to the peaceful bays and lagoons of Salvador, Brazil's coastline attracts tourists from all over the world.

RIO DE JANEIRO

Sugar Loaf Mountain, which rises above Guanabara Bay, offers a breathtaking, panoramic view of Rio de Janeiro—one of the most exciting and exotic cities in the world. Miles of white, sandy beaches and tall, graceful palm trees line the bay, and modern skyscrapers rise in the background.

Beautiful beaches

Rio de Janeiro—or simply Rio—lies nestled between forested mountains and the sparkling blue waters of the Atlantic Ocean and Guanabara Bay. Its name means *River of January,* which some historians believe refers to the month in 1503 when Portuguese seafarers first sailed into Guanabara Bay. Once the capital of Brazil, Rio is now the second largest city in South America, after São Paulo.

Rio's beautiful scenery provides an enchanting setting for its friendly, fun-loving citizens, who have been called *Cariocas* since the city's early years. This nickname may have come from a South Ameri-

can Indian expression meaning *white man's house.* Rio's long stretches of sunny beaches encourage an easygoing life style. Cariocas flock to the beaches to play volleyball and bask in the sun, or simply to enjoy the many music festivals and celebrations that take place on the sand.

Shantytowns

A world away from the glamorous beaches of Ipanema and Copacabana lies another, far less attractive side of Rio—the *favelas* (shantytowns). Here, thousands of people live in run down shacks on the hillsides and swampy shorelands.

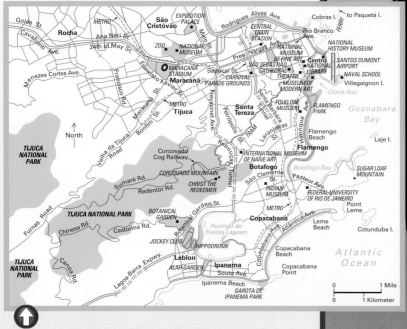

Rio de Janeiro became a major seaport in the mid-1700's, when prospectors shipped gold and diamonds to Portugal and imported many supplies. The gold trade attracted many new settlers to the city. New buildings and broad boulevards modernized Rio during the early 1900's, and today it is considered one of the most beautiful cities in the world.

A scene on Ipanema Beach, one of Rio's most glamorous resort areas, reflects two of the Cariocas' favorite activities—sunbathing and soccer. In the background, Sugar Loaf Mountain rises high above Guanabara Bay.

The name *favela* comes from a lovely wild flower that once grew on the hills, but life in these neighborhoods is far from lovely. Poverty and violent crime are major problems, and many people suffer from malnutrition. Although the government has torn down a number of slum areas and replaced them with low-cost public housing, hundreds of favelas still remain.

Samba and soccer

The lively spirit of Rio penetrates even the slums—in the form of music and dance. The driving *samba* beat echoes throughout the city, from the streets of the favelas to the elegant hotels along the Avenida Atlántica.

Samba was born as a musical form during the early 1900's when a group of young Carioca musicians began to combine traditional African rhythms with Portuguese folk songs popular at the time. It is often accompanied by the samba dance. Today, samba schools—dance clubs where dancers practice for months for the annual Carnival parade—can be found throughout the city.

The boundless energy of the Cariocas can also be seen in their love for soccer. Rio's huge Maracanã Stadium is one of the largest sports arenas in the world. A soccer game "Rio-style" is a lavish spectacle, with pounding drums, colorful flags, and firecrackers accompanying the action on the field.

Carnival

Held every year during the four days before the Christian observance of Lent, the Carnival of Rio de Janeiro symbolizes the lively spirit of the city. Although it began as a Christian festival, today's Carnival owes much of its character to African traditions, which have strongly influenced its music and dancing.

For many Cariocas, the Carnival is the most important event of the year. Preparations begin months in advance. Samba dancers and *baterias* (percussion bands) rehearse on the beach, while dressmakers create glittering, expensive costumes. Shops and street vendors stock up on colored streamers, and artists build huge *carros alegôricos* (parade floats).

When the Carnival finally begins, richly costumed Brazilians ride the floats through the boulevards of Rio, and samba schools compete for prizes in dazzling parades. Street parties, costume balls, and dancing on the beach add to the fun.

Luxurious hotels and apartment blocks overlook the crescent of Rio de Janeiro's Copacabana Beach. Situated on Guanabara Bay and the Atlantic Ocean, Rio is Brazil's chief seaport and an important center of finance, trade, and transportation.

The Carnival in Rio de Janeiro is one of the world's most colorful festivals. It is celebrated just before Lent. For four days and four nights there are parades along with dancing in the streets.

CITIES OLD AND NEW

From the modern skyscrapers of São Paulo to the cobbled streets of Ouro Prêto, each of Brazil's cities reveals another fascinating chapter of the country's history. Traveling along the nation's 4,600-mile (7,400-kilometer) Atlantic coast is a journey back to Brazil's colonial past—and a glimpse of its vision for the future.

Colonial cities

Few cities capture Brazil's early colonial days as vividly as·Recife and Olinda. Situated on the northeast coast, Recife and Olinda were occupied by Dutch colonists during their attempt to stake a claim to Brazilian territory in the 1600's. Recife is often called the *Venice of Brazil* because it is built on three rivers, and many bridges connect its islands and peninsulas.

The influence of both the Dutch—and the Portuguese who drove them out—can be seen in Recife's historic downtown area, where many lavishly decorated churches from the 1600's and 1700's still stand. The Gold Chapel of the monastery of Saint Anthony, one of the most important examples of religious art in Brazil, has a baroque design covered in gold leaf.

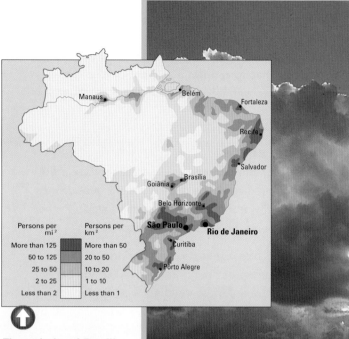

Persons per mi²	Persons per km²
More than 125	More than 50
50 to 125	20 to 50
25 to 50	10 to 20
2 to 25	1 to 10
Less than 2	Less than 1

The majority of Brazil's people live in urban areas, mainly within 100 miles of the Atlantic coast and in the southeast. Some of Brazil's first cities were built along the northeast coast during the early days of Portuguese and Dutch exploration.

Residents of a shantytown built over the river in Salvador, Bahia, cross a rickety bridge to their home. Founded in 1549, Salvador served as the capital of the Portuguese colony of Brazil until 1763.

In Olinda, now a suburb of Recife, narrow, winding streets twist up and down the hills overlooking the ocean. Many of the houses have the original latticed balconies, heavy doors, and pink stucco walls typical of the colonial period.

Farther south and hidden deep in the Brazilian Highlands is Ouro Prêto, the center of the gold and diamond trading in the 1700's. *Ouro Prêto* means *Black Gold,* and so much gold came from the hills around Ouro Prêto that the area was called *minas gerais* (general mines), which became the name for the modern state of Minas Gerais.

When the gold ran out, many people left Ouro Prêto, but the artistry of the brilliant mulatto sculptor of the period, António Francisco Lisboa da Costa (known as Aleijadinho), remains. Aleijadinho carved many human figures and church decorations out of soapstone.

Twin towers in the capital city of Brasília house congressional offices, while the Chamber of Deputies meets in the bowl-shaped structure to the far left. Viewed from the air, Brasília is laid out in a pattern that resembles a drawn bow and arrow.

São Paulo is Brazil's largest city and leading financial, commercial, and industrial center. Its many distinctive skyscrapers give the city a modern appearance.

A modern metropolis

Unlike Brazil's colonial cities with their old buildings and winding streets, São Paulo is a thoroughly modern city. Although it was founded in 1554 as an Indian mission, São Paulo has a long tradition of tearing down the old and putting up the new. Most of São Paulo's buildings are less than 100 years old, and few of the old churches remain.

Today, São Paulo is the largest city in Brazil and its leading commercial and industrial center. The wide avenues of the downtown area are lined with imaginatively designed skyscrapers and high-rise apartment blocks. Large parks and gardens provide a welcome sense of spaciousness in this crowded city.

A most contemporary capital

São Paulo is a sprawling city, growing faster every day with little direction from the government. In contrast, Brasília, the capital of Brazil, is one of the world's leading examples of large-scale city planning. Built in the east-central Brazilian wilderness in the late 1950's, Brasília is noted for its orderly development and impressive modern architecture.

The city was built as a link between the expanding south and the economically poor northeast and as a launching point for settlement of the vast interior of the country. The government hoped that Brasília would attract people from the crowded coastal cities to the underpopulated interior. Today, Brasília is a hub of highways extending north to Belem in the Amazon Region and west to Peru.

AGRICULTURE

Brazil has a vast amount of fertile farmland, and crops grown for export have been the main basis of the nation's economy since the earliest colonial days. Although factory production and service industries now contribute more to the GNP (gross national product), Brazil is still a world leader in crop and livestock production.

"Boom and bust" cycles

Beginning with the cultivation of sugar cane in the 1500's, Brazil's landowners have concentrated their efforts on growing a single crop for export in a series of "boom and bust" agricultural cycles. In each case, Brazilian farmers specialized in growing the crop that was in greatest demand on the world market at the time.

Brazil earned enormous profits during the "boom" of the crop demand, when prices were highest. But competition from other countries—or a decrease in demand—eventually brought prices tumbling down, resulting in a "bust."

In the late 1800's, for example, Brazil developed the Amazon Region's vast natural rubber resources in response to the worldwide demand for rubber products. But during the early 1900's, new rubber supplies from Asia reduced the great demand for Brazilian rubber. As rubber production decreased, coffee production increased. Then, in the 1920's, the price of coffee fell sharply, and thousands of plantation workers lost their jobs.

Crops and farming regions

During the 1980's, the Brazilian government encouraged farmers to grow a greater variety of crops. In addition to coffee, Brazil now leads all nations in growing bananas, cassava (a tropical plant with starchy roots), oranges, papayas, and sugar cane. It is also one of the world's top producers of cacao beans, cashew nuts, corn, cotton, lemons, pineapples, rice, soybeans, and tobacco.

Southern Brazil contains most of the nation's productive farmland. For many years, the state of São Paulo ranked as Brazil's chief coffee-growing region, but the northern part of Paraná eventually came to supply a larger share of the coffee crop.

Women gather the berries of the pepper plant, which will be dried and sold as peppercorns or ground into black pepper. Most rural people work on the large plantations or ranches of corporations or wealthy landowners, but some work their own small farms with traditional tools.

In addition to crops, Brazil is a leading producer of livestock, and cattle production has been a major source of wealth since World War I (1914–1918). Brazil ranks as one of the leading hog producers in the world today. Farmers also raise chickens, horses, and sheep.

In 1975, the traditional plantation crop of sugar cane became a source of fuel as well as of refined sugar. General Ernesto Geisel, who was president at the time, began a program to reduce Brazil's dependence on oil imports by substituting alcohol for gasoline made from oil. Today, a large number of Brazilian cars use alcohol distilled from sugar cane instead of gasoline.

Feeding the people

In addition to large cash crops of cacao beans, coffee, oranges, soybeans, sugar cane, and tobacco, Brazilian farmers grow such staples as cassava, beans, corn, rice, and potatoes for domestic use. However, the fast-growing population is outstripping the country's food supply. As a result, Brazil has had to import some food, particularly wheat, to feed its people.

Despite this situation, government policies—designed to reduce Brazil's huge foreign debt—encourage newly cultivated land to be used for growing high-profit cash crops. As a result, the increase in the percentage of land devoted to export crops was far greater than the increase in land used for domestic production during much of the late 1900's.

Coffee is one of Brazil's most important exports. Coffee is produced from the dried beans of the coffee plant. The drying process involves exposing trays of beans to the sun.

Brazilian laborers arrange cacao beans in an even layer to dry in the sun. Cacao beans, taken from the seed pods of an evergreen tree, are used in the production of chocolate and cocoa. Brazil is one of the world's largest suppliers of cacao.

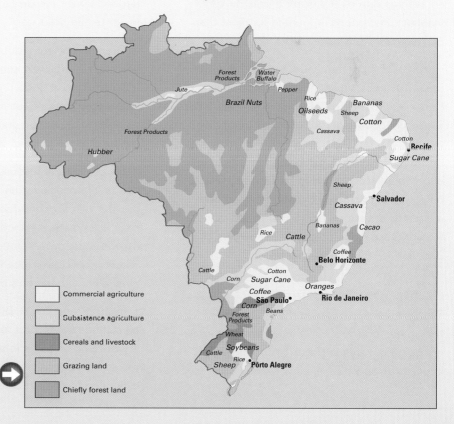

Most of Brazil's chief farming and grazing areas are centered in the south and east of the country. Agriculture accounts for about 5 percent of Brazil's economic output and employs about 20 percent of the nation's workers.

INDUSTRY

Brazil's natural resources, including its fertile farmland and rich mineral deposits, have been the backbone of the nation's economy through much of its history. However, rapid industrial growth during the mid-1900's helped Brazil become one of the world's leading manufacturing nations.

Except for textile mills, industrial development did not actually begin to blossom until Brazil became an independent nation in 1889. Before that time, Portuguese rulers had discouraged industrial development in the colony because they wanted the Brazilians to buy Portugal's manufactured goods.

Once Brazil gained independence, industry enjoyed considerable growth under the leadership of Pedro II, who ruled for almost 50 years. Textile mills, breweries, chemical plants, and glass and ceramic factories were built.

From 1948 to 1976, the nation's greatest period of industrial growth, production rose at an average rate of 9 percent a year. By 1977, manufactured goods accounted for more than 50 percent of the value of Brazil's exports.

However, such tremendous growth had its price. To speed development, the government borrowed heavily from foreign countries. Later, Brazil's huge foreign debt helped trigger hyperinflation. In 1990, President Fernando Collor de Mello froze larger bank deposits in an effort to slow down the inflation rate. He also began to sell some of the state-owned businesses to private corporations. Today, manufacturing accounts for about 20 percent of Brazil's GNP (gross national product).

Plants and factories
Today, Brazil ranks among the world's major automobile producers. Many international carmakers operate plants in Brazil. Latin America's largest iron and steel plant is located at Volta

Redonda, near Rio de Janeiro. Brazil is also a major textile producer.

Other important industries include the manufacture of airplanes, cement, chemicals, electrical equipment, food products, machinery, pharmaceuticals, paper, and transportation equipment. Most manufacturing activity is centered in the state of São Paulo.

Brazil's natural resources provide the power and raw materials for its industry. For example, the vast iron ore deposits in Minas Gerais have helped the steel industry expand. Large hydroelectric power stations on the Paraná, São Francisco, and Tocantins rivers provide almost all of the nation's electricity.

In 2007 and 2008, Brazil's state-controlled oil company, Petrobras, announced the discovery of huge oil fields in the Atlantic Ocean, off the country's southeast coast. The fields are estimated to contain billions of barrels of oil.

Technicians at an Embraer aircraft plant install the nose cone of an airplane. Despite Brazil's industrial development, millions of its people are extremely poor and work in low-paying jobs, if they can find work at all.

Service industries

From 1940 to 1980, the percentage of Brazil's workers employed in service industries increased from 20 to 40 percent. Today, service industries employ about 60 percent of Brazil's workers. Services account for a larger share of Brazil's economic output than industry and agriculture combined.

Many business services, such as banking, communications, insurance, and transportation, have developed to meet the needs of Brazil's industries. Another important area of growth has been among government agencies responsible for providing medical care and education.

Independent prospectors mine for gold at Serra Pelada (Naked Mountain) in the Amazonian state of Pará. During the dry season, which lasts from June to November, thousands of miners come here to seek their fortunes.

A truck unloads sugar cane at an ethanol plant. The ethanol produced at the plant will be used as a fuel substitute for petroleum. Sugar cane-based ethanol has made Brazil nearly independent from oil imports.

BRUNEI

Sir Muda Hassanal Bolkiah, who became sultan of Brunei in 1967, led the country through independence. He is one of the world's richest people.

Brunei (BROO ny or BROO nay), a small country in Southeast Asia, lies on the north coast of the island of Borneo. The South China Sea borders Brunei on the north, and the rest of Brunei is surrounded by Malaysia. Most of the land is flat, and the interior is heavily wooded. Brunei's climate is tropical, with average monthly temperatures of about 80° F (27° C). Rainfall averages about 100 inches (250 centimeters) a year along the coast and about 125 inches (320 centimeters) inland.

As early as the A.D. 600's, Brunei was mentioned in Chinese writings as an important trading center. The first sultan of Brunei came to power in the 1200's. During the 1400's and 1500's, Brunei was a powerful country that controlled most of the north coast of Borneo and parts of the southern Philippines. In the 1600's and 1700's, pirates used Brunei as a base for their attacks on European trading ships.

In the 1800's, in order to protect its shipping lanes between China and India, the United Kingdom took over most of northern Borneo, including Brunei. In 1888, the United Kingdom made the area a British protectorate. Brunei became an independent nation on Jan. 1, 1984.

Government

The country's official name is Negara Brunei Darussalam, which means Brunei, Abode of Peace. Its capital, Bandar Seri Begawan, is also its largest city. A monarch called a sultan heads the government and rules for life. In 1967, Sir Muda Hassanal Bolkiah was chosen as sultan. The sultan also serves as the country's prime minister, minister of finance, and minister of home affairs. Several members of his family hold high positions in the government.

FACTS

Official name:	Negara Brunei Darussalam
Capital:	Bandar Seri Begawan
Terrain:	Flat coastal plain rises to mountains in east; hilly lowland in west
Area:	2,226 mi² (5,765 km²)
Climate:	Tropical; hot, humid, rainy
Main rivers:	Belait, Tutong, Temburong
Highest elevation:	Pagon, 5,932 ft (1,808 m)
Lowest elevation:	South China Sea, sea level
Form of government:	Monarchy ruled by a sultan
Head of state:	Sultan and prime minister
Head of government:	Sultan and prime minister
Administrative areas:	4 daerah-daerah (districts)
Legislature:	Majlis Masyuarat Megeri (Legislative Council)
Court system:	Supreme Court
Armed forces:	7,000 troops
National holiday:	National Day - February 23 (1984)
Estimated 2010 population:	411,000
Population density:	185 persons per mi² (71 per km²)
Population distribution:	74% urban, 26% rural
Life expectancy in years:	Male, 73; female, 78
Doctors per 1,000 people:	1.1
Birth rate per 1,000:	18
Death rate per 1,000:	3
Infant mortality:	10 deaths per 1,000 live births
Age structure:	0-14: 29%; 15-64: 68%; 65 and over: 3%
Internet users per 100 people:	50
Internet code:	.bn
Languages spoken:	Malay (official), English, Chinese
Religions:	Muslim (official) 67%, Buddhist 13%, Christian 10%, other 10%
Currency:	Bruneian dollar
Gross domestic product (GDP) in 2008:	$14.40 billion U.S.
Real annual growth rate (2008):	0.4%
GDP per capita (2008):	$36,275 U.S.
Goods exported:	Clothing, crude oil, natural gas
Goods imported:	Food, machinery, manufactured goods, transportation equipment
Trading partners:	Indonesia, Japan, Malaysia, Singapore, South Korea, United States

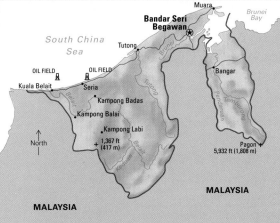

Muara

Bandar Seri Begawan

South China Sea

Tutong

Brunei Bay

OIL FIELD OIL FIELD

Kuala Belait

Seria

Bangar

Kampong Badas

Kampong Balai

Kampong Labi

North

1,367 ft (417 m)

Pagon
5,932 ft (1,808 m)

MALAYSIA

MALAYSIA

0   20 Miles
0   20 Kilometers

BRUNEI

Resources

The discovery of oil off the coast in 1929 brought great wealth to Brunei. The income from oil exports has enabled the people of Brunei to enjoy a high standard of living. However, the nation's petroleum and gas reserves are expected to run out eventually.

Petroleum, petroleum products, and the natural gas often found with petroleum account for almost all of Brunei's exports. However, the petroleum and gas industry employs only about 5 percent of the country's labor force. The government is by far the nation's largest employer. Agriculture plays a small role in Brunei's economy.

People

About three-fourths of Brunei's people live in urban areas, and about one-fourth live in rural areas. About two-thirds of the people are Malays, and most are Muslims—followers of the faith of Islam. The Chinese, the largest minority, make up about 11 percent of the population. Most of the Chinese are Christians, and a small percentage are Buddhists. Most Bruneians speak Malay, the official language, but English and Chinese are also used.

In urban areas, most Bruneians wear Western-style clothing, but many Muslim women wear outfits of long skirts and long-sleeved blouses. In rural areas, many men and women wear loose shirts and *sarongs*, long pieces of cloth worn as a skirt and tied at the waist. Most city dwellers live in modern houses or apartment buildings made of brick or stone, while most rural homes are wooden and have thatched roofs.

The government provides free schooling and medical services for its people. Most Bruneian children complete elementary school, and many go on to high school. The nation's first university, the University of Brunei Darussalam, opened in 1985. Many Bruneians study at foreign universities, and the government pays for their education.

BULGARIA

Bulgaria lies on the Balkan Peninsula in southeastern Europe. It is bordered to the north by Romania, to the west by Serbia and Macedonia, and to the south by Greece and Turkey. The Black Sea lies to the east.

A people called the Thracians established the first civilization in what is now Bulgaria about 3,000 years ago. The region became part of the Roman Empire during the A.D. 40's. In later years, Bulgaria twice ruled a powerful kingdom that covered most of the Balkans. In the late 1300's, the Ottoman Empire conquered the country. Bulgaria became independent in 1908 and came under Communist rule in 1946. Like other Balkan countries, the nation experienced dramatic political upheaval during the late 1980's and early 1990's, as Communist governments declined.

During the years under Communist control, Bulgaria had changed from an agricultural country to an industrialized country. This shift brought the nation some capital and international credit, but conditions for the working people remained poor. Wages were low, and food, housing, and consumer goods were in short supply.

In late 1989, human-rights activists and environmentalists challenged the dominance of Bulgaria's Communist Party in the largest demonstrations since the end of World War II (1939-1945). Only a few days later, Todor Zhivkov, who had been head of state and leader of the Bulgarian Communist Party for the previous 27 years, was removed from office. Further demonstrations calling for democratic reforms and free elections were held in the capital city of Sofia.

In early 1990, the Communist Party gave up its monopoly of power, and the party changed its name to the Bulgarian Socialist Party (BSP). In June, Bulgaria held its first free, multiparty elections in 44 years. That August, Zhelyu Zhelev of the Union of Democratic Forces (UDF)—a coalition of 16 opposition parties—became Bulgaria's first non-Communist head of state since 1944. Bulgaria adopted

FACTS

Official name:	Republic of Bulgaria
Capital:	Sofia
Terrain:	Mostly mountains with lowlands in north and southeast
Area:	42,823 mi² (110,910 km²)
Climate:	Temperate; cold, damp winters; hot, dry summers
Main rivers:	Danube, Iskŭr, Maritsa, Ogosta, Tundzha
Highest elevation:	Musala Peak, 9,596 ft (2,925 m)
Lowest elevation:	Black Sea, sea level
Form of government:	Parliamentary democracy
Head of state:	President
Head of government:	Chairman of the Council of Ministers (Prime minister)
Administrative areas:	28 oblasti (provinces)
Legislature:	Narodno Sobranie (National Assembly) with 240 members serving four-year terms
Court system:	Supreme Administrative Court, Supreme Court of Cassation, Constitutional Court
Armed forces:	40,700 troops
National holiday:	Liberation Day - March 3 (1878)
Estimated 2010 population:	7,503,000
Population density:	175 persons per mi² (68 per km²)
Population distribution:	71% urban, 29% rural
Life expectancy in years:	Male, 69; female, 76
Doctors per 1,000 people:	0.3
Birth rate per 1,000:	10
Death rate per 1,000:	15
Infant mortality:	10 deaths per 1,000 live births
Age structure:	0-14: 13%; 15-64: 70%; 65 and over: 17%
Internet users per 100 people:	31
Internet code:	.bg
Languages spoken:	Bulgarian, Turkish, Roma
Religions:	Bulgarian Orthodox 82.6%, Muslim 12.2%, other Christian 1.2%, other 4%
Currency:	Lev
Gross domestic product (GDP) in 2008:	$50.44 billion U.S.
Real annual growth rate (2008):	6.0%
GDP per capita (2008):	$6,682 U.S.
Goods exported:	Clothing, copper and copper products, footwear, iron and steel, petroleum products
Goods imported:	Crude oil and natural gas, machinery, metals and ores, transportation equipment
Trading partners:	Germany, Greece, Italy, Russia, Turkey

Once part of the Roman Empire, Bulgaria, became a kingdom in 681. The country was ruled by the Byzantines between 1018 and 1186. Bulgaria then kept its independence until 1396, when it was conquered by the Ottomans.

The Alexander Nevsky Cathedral in Sofia is the most familiar symbol of Bulgarian religious life. The structure is one of the largest Eastern Orthodox cathedrals in the world. The church was built in the neo-Byzantine style from 1904 to 1912.

a new constitution in July 1991. The Constitution made the prime minister the most powerful government official. Bulgaria joined the European Union in 2007.

About 85 percent of the country's people are of Bulgarian ancestry, descended from Slavs and Bulgars. Bulgarian, the country's official language, is related to Russian and other Slavic languages and written in the Cyrillic alphabet.

Turks make up Bulgaria's largest ethnic minority and represent about 10 percent of the population. Because Bulgaria suffered as part of the Ottoman Empire from the 1300's to the end of the 1800's, many Bulgarians resent the Turks.

In 1984 and 1985, the government tried to force the Turks to adopt Bulgarian names, and it banned the use of the Turkish language in public. Some who ignored these new rules were killed by Bulgarian troops. In 1989, about 344,000 Turkish-speaking Bulgarians sought refuge in Turkey, but nearly one-third of them returned to Bulgaria after a few months. In 1990, the Bulgarian government ended its anti-Turkish program, but some ethnic disturbances continued.

ENVIRONMENT

Bulgaria is a mountainous land broken by fertile valleys and plains. The country's four main land regions are the Balkan Mountains, the Danubian Plateau, the Transitional Mountains and Lowlands, and the Rhodope Mountains. Because of the sharp contrasts in its terrain, Bulgaria's climate varies greatly from region to region.

The country's landscape is shaped chiefly by the Balkan Mountains, which stretch across Bulgaria from west to east, dividing the country in half. The northern half—the Danubian Plateau—has cold winters but warm and humid summers. The southern half—the Transitional Mountains and Lowlands—has cool winters and hot, dry summers.

In the mountain regions, weather conditions change yet again, depending on the altitude and the distance from the sea. The country's average rainfall is 25 inches (63 centimeters), and snowfall is light, except in the mountains. Near the Black Sea coast, winters are mild and summers are hot.

Bulgaria's coast extends about 175 miles (282 kilometers) along the Black Sea. The nation's sandy beaches, sunny climate, and fascinating historical ruins draw vacationers from all over Europe. Bulgaria's leading Black Sea ports are Varna and Burgas.

Northern Bulgaria

The Danubian Plateau, which covers northern Bulgaria from the Danube River south to the Balkan Mountains, is a vast sheet of limestone covered with river silt. As a result, this region has the country's most fertile farmland. The Danube River forms most of the border between Bulgaria and Romania. Several of the Danube's tributaries—including the Iskŭr and the Yantra—flow northward from the mountains.

The Danubian Plateau is partly flat, becoming more rolling as it approaches the foothills of the Balkans. In the valleys and flatter areas, where the climate is more humid, fruit, vegetables, and wine grapes flourish. The drier uplands are used for growing corn and wheat. In the northeast, the vast expanses of scrub grassland provide ideal pastures for sheep.

Snow-capped peaks rise above the Rila
Monastery, hidden away in the Rila
Mountains of southwestern Bulgaria.
Bulgaria's highest peak, Musala, rises
9,596 feet (2,925 meters) along the
northern tip of the Rhodope Mountains.

A small town occupies a sheltered valley in
southeastern Bulgaria. The Maritsa and
Tundzha river valleys support the cultivation
of fruits and vegetables, such as apples,
grapes, pears, tomatoes, and watermelons.

The mountainous south

The Rhodope Mountains stretch across southern Bulgaria, forming a natural boundary with Greece. Evergreen forests and alpine meadows are a dominant feature of this landscape. The country's highest point, Musala Peak, rises 9,596 feet (2,925 meters) above sea level in the Rhodope Mountains.

Northwest of the Rhodope, in the Rila Mountains, stands the Rila Monastery, a historic shrine revered by all Bulgarians and now the site of a national museum. West of the Rhodope rise the peaks of the Pirin Mountains, where bears, wolves, and wildcats are found among the abundant wildlife.

On the Black Sea coast, sandy beaches are interspersed with dunes, rocky cliffs, and forest-covered hills. The large coastal resorts of Varna and Burgas are known for their splendid beaches and parks. Broad expanses of reeds and water lilies mark the deltas of the Kamchiya and Ropotamo rivers, and farther north, monk seals bask on the rocks along the shore.

Most of the Balkan Mountains are not very high. The taller peaks lie mostly in the west, and some tower about 7,120 feet (2,170 meters) along the border with Serbia. Generally, the Balkans form lengthy ridges and small plateaus where sheep graze and forests grow. Many mountain passes allow traffic to flow easily between the Danubian Plateau and the mountain regions of southern Bulgaria.

Immediately to the south of the Balkan Mountains lie a number of lower mountain chains, known as the Sredna Gora. A zone of fertile basins, where roses and wine grapes are cultivated, lies between the Balkans and these lower chains. Bulgaria's capital city of Sofia is situated in one of these basins, at an altitude of 1,800 feet (550 meters).

Rows of corn grow on the Danubian Plateau, in the
region's fertile soil. The crops are aided by the humid
summer conditions. In the southern part of Bulgaria,
where the climate is drier, irrigation provides moisture
for crops.

ECONOMY

Following the Communist take-over in 1946, Bulgaria changed from an economy dominated by privately owned agricultural interests to one in which state-owned industry was of primary importance. During the period of industrialization, heavy industry was developed rapidly, and extremely high production goals were set by the government.

However, poor management and shortages of fuel and skilled labor have slowed economic growth in Bulgaria. At the same time, wherever an industry was allowed some degree of self-administration or profit-making, efficiency and productivity improved.

During the period of Communist rule, Bulgaria maintained exceptionally close ties with the Soviet Union, partly out of gratitude for Russia's help in 1878 when they gained freedom from Turkish rule. Because of its relationship to the Soviet Union, Bulgaria was given the responsibility for research, development, and production in the field of microelectronics for all Eastern bloc countries. Bulgaria's historic city of Plovdiv became the nation's center for high technology.

Beginning in 1985, Bulgaria's trade links with Eastern bloc countries were gradually abandoned, and Bulgaria began to seek greater cooperation with Western countries. However, the rigidly controlled and heavily subsidized economy of Bulgaria was unable to compete with the free markets of Western countries and corporations.

In 1989, the Bulgarian government announced plans to reform industry and agriculture. These reforms called for a decrease in centralized planning, permitting the establishment of small, private enterprises. Some foreign participation was to be allowed in these enterprises, and shares in newly formed companies were to be made available to foreign, as well as Bulgarian, investors.

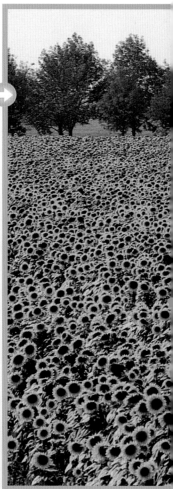

A field of golden sunflowers outside the town of Lom, in northwestern Bulgaria, is almost ready for harvest. The seeds of the sunflowers are crushed to extract a high-quality vegetable oil.

Tank trucks await refueling at an oil storage depot at Varna, one of the country's leading ports on the Black Sea. Although Bulgaria has two oil fields, its reserves are very small, and the country imports most of its fuel.

Bulgaria's Black Sea coast attracts millions of vacationers, primarily from Eastern European countries. Tourists who come to enjoy the coast's historic sites, mild climate, and sandy beaches have greatly benefited the Bulgarian economy.

A cooper (barrel maker) produces wooden barrels for a local vineyard. Bulgaria produces red and white wines, as well as brandies, fruit liqueurs, and even a spirit made from roses. Some wine is exported to Western countries.

Roses are harvested by farmers in central Bulgaria for their essence —a fragrant oil made by distilling rose petals plucked from still-closed buds. Thousands of buds are harvested by hand to produce 2 pints (1 liter) of oil.

Industry and energy

Today, manufacturing, mining, and energy production account for about half of Bulgaria's *net material product* (NMP)—the total value of goods, and of services used in the production of these goods, by the country in a year. This sector also employs about a third of the nation's workers. Bulgaria's major industrial centers are Sofia, Dimitrovgrad, Plovdiv, Ruse, and Varna. The top manufacturing industries produce chemicals, machinery, metal products, processed foods, and textiles.

Bulgaria has small deposits of many kinds of minerals. The country mines coal, copper, kaolin, lead, pyrite, salt, sulfur, and zinc. A nuclear power station at Kozloduy produces much of the country's electricity. Bulgaria must import most of its fuel.

Agriculture

Despite the industrialization of Bulgaria, agriculture still accounts for about 20 percent of the country's net material product and employs about 25 percent of the country's workers. Farmland covers about 15 million acres (6 million hectares), or more than half of Bulgaria.

Bulgaria's chief farm product is grain. Wheat and corn are the leading crops, and other grains include barley, oats, rice, and rye. Bulgarian farmers also grow a wide variety of fruits and vegetables, including apples, grapes, potatoes, pears, sugar beets, tomatoes, and watermelons. Roses are grown for their sweet-smelling oil, which is used as the basis for many world-famous perfumes. Livestock production, including dairy and beef cattle, chickens, pigs, and goats, is also an important activity. Cow's milk is a major farm product.

BURKINA FASO

The country of Burkina Faso *(bur KEE nuh FAH soh)* in western Africa was known as Upper Volta until 1984. That year, the government changed the name to *Burkina Faso,* which means *land of the honest people.* It lies on the western bulge of Africa, about 600 miles (970 kilometers) east of the Atlantic Ocean. The country has about 15 million people.

Landlocked Burkina Faso is one of the least developed countries on the African continent. This dry, rocky plateau turns green for only a few months each year, because its thin soil does not hold rain water well. Most of the rainfall quickly runs off into the country's many rivers. Because the country has poor soil and no mineral deposits, its people have only the bare necessities of life. Most make their living by raising cattle or farming.

Cattle raising is an important economic activity in Burkina Faso. In the river valleys, farmers raise such food crops as beans, corn, millet, rice, sorghum, and *fonio* (a grass whose seeds are used as a cereal). Cash crops include cotton, peanuts, and *shea nuts* (seeds that contain a fat used to make soap).

Because Burkina Faso lacks resources, many of its young men work in neighboring countries on plantations or as city laborers. The money they send home is important to the nation's income as well as to their families.

Most of the people of Burkina Faso belong to one of two major cultural groups—the Voltaic and the Mande. The Voltaic group includes the Mossi, the Bobo, the Gurunsi, and the Lobi peoples. The Mande group includes the Boussance, the Marka, the Samo, and the Senufo peoples.

The Voltaic Mossi, who make up about half the country's population, are mainly farmers who live in the central and eastern parts of the country. The typical Mossi family lives in a *yiri,* a group of mud huts built around a small court,

FACTS

Official name:	Burkina Faso
Capital:	Ouagadougou
Terrain:	Mostly flat to dissected, undulating plains; hills in west and southeast
Area:	105,869 mi² (274,200 km²)
Climate:	Tropical; warm, dry winters; hot, wet summers
Main rivers:	Black Volta, Red Volta, White Volta
Highest elevation:	Aiguille de Sindou, 2,352 ft (717 m)
Lowest elevation:	Black Volta River, 656 ft (200 m)
Form of government:	Parliamentary republic
Head of state:	President
Head of government:	Prime minister
Administrative areas:	45 provinces
Legislature:	Assemblee Nationale (National Assembly) with 111 members serving five-year terms
Court system:	Supreme Court, Appeals Court
Armed forces:	10,800 troops
National holiday:	Republic Day - December 11 (1958)
Estimated 2010 population:	15,454,000
Population density:	146 persons per mi² (56 per km²)
Population distribution:	81% rural, 19% urban
Life expectancy in years:	Male, 50; female, 53
Doctors per 1,000 people:	Less than 0.05
Birth rate per 1,000:	44
Death rate per 1,000:	14
Infant mortality:	89 deaths per 1,000 live births
Age structure:	0-14: 46%; 15-64: 51%; 65 and over: 3%
Internet users per 100 people:	0.9
Internet code:	.bf
Languages spoken:	French (official), native African languages belonging to Sudanic family
Religions:	Muslim 61%, Christian (mainly Roman Catholic) 23%, indigenous beliefs 15%, other 1%
Currency:	Communaute Financiere Africaine franc
Gross domestic product (GDP) in 2008:	$8.10 billion U.S.
Real annual growth rate (2008):	4.5%
GDP per capita (2008):	$561 U.S.
Goods exported:	Cotton, livestock, other agricultural products
Goods imported:	Food, machinery, petroleum, vehicles
Trading partners:	Côte d'Ivoire, France, Ghana, Togo

where the family's sheep and goats are kept. For more than 800 years, the Mossi have had a kingdom in the region of Burkina Faso headed by the *Moro Naba,* or Mossi chief. A Moro Naba still holds court in the city of Ouagadougou.

The Voltaic Bobo, Gurunsi, and Lobi each make up less than 10 percent of the population. The Bobo live in large villages in southwestern Burkina Faso. The Gurunsi live around the city of Koudougou and have adopted modern changes fairly readily. The Lobi live in the Gaoua region and have turned from hunting and farming to migrant labor.

The Mande Boussance, Marka, Samo, and Senufo peoples are branches of Mande groups living in the neighboring countries of Mali, Guinea, and northern Cote d'Ivoire.

In addition to the Voltaic and Mande peoples, several hundred thousand Fulani and Tuareg live in Burkina Faso. These nomads roam the pastures in the north of the country with their goats, sheep, and other livestock.

The Mossi have the longest history among the people of Burkina Faso, but their kingdom was not discovered by Europeans until the 1800's. In 1897, France captured Ouagadougou. In 1919, France created the colony of Upper Volta in the region.

On August 5, 1960, Upper Volta became an independent republic. Since then, several civilian governments have been overthrown by army officers, who in turn have been overthrown in other military coups. Today, the president is Burkina Faso's most powerful official. A Council of Ministers helps the president carry out government operations. A legislature called the National Assembly makes the laws.

Captain Blaise Compaoré, who came to power in a military *coup* (revolt against the government) in 1987, was elected president in 1991. He was reelected in 1998, 2005, and 2010.

A Bobo villager takes a break from her chores in the shade of thatch-roofed granaries. The Bobo people live in large villages in southwestern Burkina Faso, where they build castlelike structures with clay bricks and straw.

Burkina Faso lies about 600 miles (970 kilometers) east of the Atlantic Ocean on the western bulge of Africa. Consisting mostly of a wooded and grassy plateau, it is one of Africa's poorest nations.

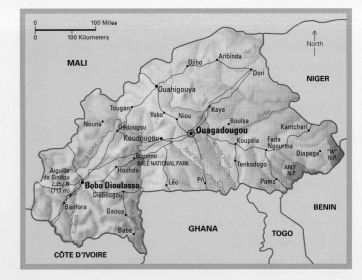

BURUNDI

Burundi *(buh RUHN dee* or *bu ROON dee),* in east-central Africa, is one of the continent's smallest and most crowded countries. The nation has more than 9 million people, with an average of 876 people per square mile (338 people per square kilometer).

Burundi has few minerals and little industry—most of the people are farmers who can raise only enough food to feed their families. The main crops are bananas, beans, cassava, corn, and sweet potatoes. Some farmers also raise cattle and other livestock. In addition, Lake Tanganyika provides a major source of fish.

A large majority of the people of Burundi belong to the Hutu ethnic group—mainly farmers who are poor. The Tutsi group, while a minority, dominates the nation politically and economically. It is not known when the Hutu arrived, but they were living there when the Tutsi invaded from the north.

Most of the remaining population are Twa people, a Pygmy group. The Twa once hunted and gathered wild food, such as berries, but today many Twa make pottery and farm for a living. The Twa were probably the first inhabitants of the area and may have lived there since prehistoric times.

The Tutsi, who were more powerful than the Hutu, agreed to protect the Hutu if the Hutu would raise crops for them. However, the region was actually ruled by a small group, called the *Ganwa,* who ruled both the Tutsi and the Hutu and became wealthy. The people's king, called *mwami,* had little power.

In 1897, the Germans conquered the area that is now Burundi and Rwanda, the country to the north. Belgium later took control of the region, then called Ruanda-Urundi. In 1961, Urundi voted to become the independent monarchy of Burundi, while Ruanda voted to become the republic of Rwanda. The two became independent on July 1, 1962. By then, the power of the Ganwa had ended, and the Tutsi controlled Burundi.

FACTS

Official name:	Republika y'u Burundi (Republic of Burundi)
Capital:	Bujumbura
Terrain:	Hilly and mountainous, dropping to a plateau in east, some plains
Area:	10,747 mi² (27,834 km²)
Climate:	Equatorial; generally moderate; wet seasons from February to May and September to November with average yearly rainfall of about 58 in. (147 cm)
Main rivers:	Rusizi, Ruvironza, Ruvubu
Highest elevation:	Mount Heha, 8,760 ft (2,670 m)
Lowest elevation:	Lake Tanganyika, 2,533 ft (772 m)
Form of government:	Republic
Head of state:	President
Head of government:	President
Administrative areas:	16 provinces
Legislature:	Parliament consisting of the Senate with 54 members and the Assemblee Nationale (National Assembly) with at least 100 members serving five-year terms
Court system:	Cour Supreme (Supreme Court), Constitutional Court
Armed forces:	20,000 troops
National holiday:	Independence Day - July 1 (1962)
Estimated 2010 population:	9,417,000
Population density:	876 persons per mi² (338 per km²)
Population distribution:	90% rural, 10% urban
Life expectancy in years:	Male, 49; female, 51
Doctors per 1,000 people:	Less than 0.05
Birth rate per 1,000:	46
Death rate per 1,000:	16
Infant mortality:	107 deaths per 1,000 live births
Age structure:	0-14: 45%; 15-64: 52%; 65 and over: 3%
Internet users per 100 people:	0.8
Internet code:	.bi
Languages spoken:	Kirundi (official), French (official), Swahili
Religions:	Roman Catholic 62%, indigenous beliefs 23%, Muslim 10%, Protestant 5%
Currency:	Burundi franc
Gross domestic product (GDP) in 2008:	$1.05 billion U.S.
Real annual growth rate (2008):	4.5%
GDP per capita (2008):	$126 U.S.
Goods exported:	Coffee, cotton, sugar, tea
Goods imported:	Food, petroleum products, vehicles
Trading partners:	Belgium, Kenya, Pakistan, Saudi Arabia, Uganda

After independence, political unrest and ethnic conflict troubled Burundi. The Hutu resented the power of the Tutsi. Political assassinations took place in 1965, and the king was overthrown in a military revolt in 1966. Then in 1972, the Hutu revolted. By the end of their unsuccessful rebellion, about 100,000 people, most of them Hutu, had died. A second army revolt in 1976 made Colonel Jean-Baptiste Bagaza president.

Bagaza restricted the Roman Catholic Church, which supported the rights of the Hutu. He was overthrown and replaced in 1987, but Hutu-Tutsi violence again broke out in 1988. Thousands of people died in the conflict or fled the nation.

In April 1994, Burundi's president was killed, along with Rwanda's president, in a suspicious airplane crash at an airport in the Rwandan capital of Kigali. His replacement was overthrown in 1996. Outbreaks of ethnic violence occurred sporadically throughout the middle to late 1990's. Over 300,000 people were killed in these conflicts.

In 2004, many of Burundi's political groups signed a power-sharing agreement. In February 2005, voters approved a new constitution that guarantees both the Hutu and the Tutsi a share in a variety of government offices. The last Hutu rebel group disarmed in 2009 and became a political party.

Burundi is a small, landlocked country whose location makes overseas trade difficult and expensive. Because the nation is situated far inland, goods must be loaded and unloaded from lake ships and railroad cars many times before reaching their destination.

Village women purify their water by pouring it through a gourd containing a purifying agent. Impure drinking water has caused illness in Burundi. A government program designed to improve the rural water supply began in 1986.

An elderly Tutsi man rests in the shade. Many Tutsi own livestock. Although a minority, the Tutsi hold most of Burundi's wealth and control its government and army. They are sometimes called the Watusi.

On the highlands that cover much of Burundi, farmers have cleared the once-wooded plateaus. Poor farming methods and heavy rains have eroded once-fertile volcanic soils in the west.